TOUCH WOOD

TOUCH WOOD

BC Forests at the Crossroads

Harbour Publishing

HARBOUR PUBLISHING CO. LTD.
P.O. Box 219
Madeira Park, BC
Canada VON 2HO

Edited by Laurel Bernard
Statistical graphics prepared by Michael Travers
Cover painting by Gord Halloran
Cover design by Roger Handling
Drawings on pages 104–107 and 130–131 by Jim Brennan
Type and page composition by Vancouver Desktop Publishing Centre
Printed and bound in Canada

Canadian Cataloguing In Publication Data

Main entry under title:
  Touch wood

  Includes bibliographical references and index.
  ISBN 1-55017-074-0

1. Forests and forestry—British Columbia.
2. Forest policy—British Columbia.   I. Drushka, Ken.
II. Nixon, Bob.   III. Travers, Ray.
SD 568.B7T68 1993   333.75'09711   C92- 091510-8

Contents

Foreword

Ken Drushka

TO UNDERSTAND THE GENESIS of this book it helps to revisit the context in which it was conceived. In May 1991, several of the contributors met in Victoria to discuss the future course of British Columbia's forest policy debate. This was a few weeks after the Forest Resource Commission (FRC), chaired by Sandy Peel, had released its report; and a few months before the provincial election that, even then, was clearly in the bag for the New Democratic Party.

During the previous two years British Columbians had engaged in an unprecedented public debate on forest policy, touched off by Social Credit Forest Minister Dave Parker's ill-fated attempt to enshrine the status quo in an expanded Tree Farm Licence system. The subsequent hearings of the FRC produced 1700 formal submissions involving the energies of thousands of citizens from all walks of life. Never before in the history of the province had so many people become involved in what is arguably the most important public policy issue in BC. By comparison, the seminal Pearse Report of 1976 was based on fewer than 200 submissions.

At the time the FRC report was released, the province was primed for drastic forest policy reform. In its analysis, the Commission's report did not flinch. It was a devastating indictment of existing forest administration and reflected the public mood. But its recommendations for change were disappointing and elicited almost no support from industry people, environmentalists, politicians of any stripe or the general public.

The lame-duck Socred government acted as if the FRC had never existed, and the incoming NDP administration chose to ignore it and abolish the commission. It was clear that the expectations of all those who had participated in the debate over the past two years were not going to be met. The disappointment felt by many of us at that point was not so much that our particular concerns or interests were not being dealt with, but that the entire process of the previous years was not going to lead to any resolution of the various conflicts in the woods.

Most of the participants in our May 1991 discussion knew that the Harcourt government and its Forest Minister-in-waiting, Dan Miller, were not going to rock any boats. We realized that without the focus of the FRC, public debate would cease and interest would wane. The greatest opportunity ever to reform forest policy on the basis of informed public involvement was being lost.

This book is a consequence of that realization. We left that meeting with a list of potential contributors who were approached in subsequent weeks. Those who were able to participate shared one assumption: BC forest policy is seriously flawed and serious reforms are needed. Since we began the process of producing this book two years ago, little has changed. Public debate has been replaced by a series of government-created forums or tables, such as the Commission on Resources and the Environment, in which various stakeholders are allowed to participate. The Cabinet decision on the future of Clayoquot Sound forests suggests there has been little improvement in the planning and conflict resolution processes. There are no indications the current government intends to confront the major issue of tenure, as recommended by Pearse in 1976, by the FRC and, most recently, by its own Select Standing Committee report on the lumber remanufacturing sector. The Harcourt government's major forest-sector initiative, bringing Allowable Annual Cuts down to more realistic levels, is being done in a policy vacuum that is leading to the elimination of thousands of jobs and generating anger and despair throughout the province.

With rare exceptions, the contributors to this effort have no institutional affiliations to industry, environmental organizations or government. But all of them have deep roots and a wide range of experience in the province's forests, the forest industry, government and academic institutions. There are still major decisions to be made about the management of BC's forests; a provincial election will be on us before we know it and, as always, forestry will be one of the issues.

The authors all believe in the critical importance of ongoing public participation in the debate on forest policy. The purpose of the book is not to provide answers, but to stimulate discussion.

In addition to those who have written the papers in this book, thanks are also due to Laurel Bernard for her patience and perseverance in editing it, and to Howard White, whose understanding of its need has made it possible.

Ken Drushka
July 1993

Chronology

Significant Events in BC Forest Policy

Ray Travers

1793 Alexander Mackenzie arrives in Dean Channel on July 22— "From Canada by Land."

1807 David Thompson sees the Upper Columbia. In 1811, he journeys down the Columbia River to its mouth.

1808 Simon Fraser reaches the Fraser River delta.

1818 Joint occupation of the Oregon Territory by Great Britain and the United States.

1843 Hudson Bay headquarters relocated at Fort Victoria.

1849 Crown Colony of Vancouver Island created.

1858 Crown Colony of British Columbia created.

1865 Land Ordinance passed, marking the beginnings of a land policy.

1869 Gold Mining Act passed, to regulate operations of free miners.

1871 British Columbia joins Confederation.

1875-1899 Era of railway speculation in forest land.

1887 Esquimalt and Nanaimo Railway Land Grant and the Peace River Block created.

1891 Land Act passed introducing a classification of "lands suitable for lumbering."

1905 Public ownership of forest land becomes official forest policy; forests opened up for granting of temporary tenures.

1906 Timber Manufacture Act passed banning log exports from future crown grants.

1907 Government stops granting more temporary tenures; Dr. Bernhard Fernow conducts study of timber inventory.

1910   Royal Commission of Inquiry publishes the Fulton Report.

1912   First Forest Act passed. H.R. MacMillan becomes first Chief Forester. Establishment of Forest Reserves authorized. Establishment of Timber Sale License Tenure to be awarded by public auction.

1918   H. N. Whitford and R. D. Craig publish the first inventory, "Forests of British Columbia," for the Dominion Commission on Conservation.

1937   F. D. Mulholland publishes updated forest inventory, "The Forest Resource of British Columbia."

1942   Chief Forester C.D. Orchard delivers rationale for sustained yield forest management to A. Wells Gray, Minister of Lands: "Forest Working Circles: An Analysis of Forest Legislation in British Columbia as it Relates to Disposal of Crown Timber, and Proposed Legislation to Institute Managed Harvesting on the Basis of Perpetual Yield."

1945   Royal Commission of Inquiry publishes the Sloan Report, "The Forest Resources of British Columbia."

1947   Forest Act revised authorizing granting of Forest Management Licenses.

1956   Royal Commission of Inquiry publishes the second Sloan Report, "The Forest Resources of British Columbia."

1958   First Continuous Forest Inventory published; Forest Act amended.

1976   Royal Commission on Forest Resources publishes the Pearse Report, "Timber Rights and Forest Policy."

1978   Forest Act amended consolidating tenure rights in legislation.

1991   Forest Resources Commission publishes the Peel Report, "The Future of Our Forests."

1992   Commission on Resources and Environment created, Commissioner Stephen Owen.

Forest Tenure

Forest Ownership and the Case for Diversification

Ken Drushka

NINETEEN NINETY-TWO MARKED THE 50TH ANNIVERSARY of possibly the most significant event in the evolution of British Columbia forest policy. The fact that it occurred out of public view and, even to this day, remains largely unacknowledged says almost as much about the determination of forest policy in this province as does the event itself.

On August 27, 1942, the province's Chief Forester, C.D. Orchard, sent a confidential memo to A. Wells Gray, Minister of Lands in the coalition government, outlining proposals for a radical change in provincial forest policy. There were two significant components in this proposal: the adoption of sustained-yield forestry on provincially owned forest lands, and the creation of sustained-yield management units that would leave land ownership in Crown hands.

For more than a year the document was circulated discreetly at Cabinet level and among Premier John Hart's friends in the industry. The government then decided to implement Orchard's plan but, to forestall the anticipated opposition, set up a Royal Commission on Forestry in December, 1943. Orchard presented the concept outlined in his memo to Chief Justice Gordon Sloan, the commissioner, who dutifully included them in his recommendations. On April 3, 1947, a new Forest Act, incorporating Orchard's proposals, was passed in the Legislature.

During the course of these events, there was considerable public debate, but most of it concerned details of implementation and not the fundamental assumptions underlying the proposals themselves. One principle not discussed, except within a very small circle, was the issue of public versus private ownership of forest land. When the general debate ended with passage of the new Forest Act in 1947, the question of forest ownership was put to rest and has only recently come back into public view. One of the more curious features of BC's forest history is that an open debate on the relative merits of public and private forest land ownership has never been held. If anything, it is a debate that has been actively discouraged by foresters, resource technocrats, politicians, academics and corporate managers. The conventional response to suggestions that increased private ownership of forest land merits consideration is that the public will not stand for it. The public, of course, has never been asked what it thinks of the idea.

In his 1976 Royal Commission report, Peter Pearse discussed the pros and cons of private ownership. Retention of state ownership, he wrote, offered two important benefits: it allows the government to protect and enhance the values of forest land that does not produce financial gains to its owners, and it provides the government with a powerful tool to shape and direct economic development:

> It is primarily for these two reasons that I recommend no change in the general policy of retaining Crown title to unalienated forest land. Moreover, judging from the evidence presented at my public hearings, the forest industry is content to rely primarily on well-designed contractual rights to Crown timber, and the public at large favours Crown ownership. Finally, the growing competence of the public agencies responsible for management of Crown resources, and the evidence of good management on many licensed lands, demonstrate that effective resource management does not require private ownership.

The most recent expression of what is by now a reflexive resistance to the question of increased private ownership can be found in the April 1991 report of the Forest Resources Commission (FRC). In this report the reluctance—in fact, the refusal—to consider the alternative of private ownership is curious because of its lack of supporting argument.

The Peel Commission correctly identifies the tenure structure recommended in Orchard's secret memo as one of the major weaknesses in a forest policy framework that has failed to achieve either its economic or ecological objectives. In glowing terms it describes the diversified land ownership patterns of other strong forest economies, and concludes:

> Other jurisdictions rely much more on principles of private ownership and contractual relationships to guarantee enhanced stewardship and secure wood supplies. The Commission believes that the "self-interest" represented by these principles of private ownership should lead to better stewardship and a better representation of other values.

Then, after several pages building up to the issue which forms one of the foundation stones of provincial forest policy, the FRC dismisses the option of private ownership in one short sentence:

> The Commission has concluded that selling off the province's public lands is not a realistic alternative.

That's it. There is no analysis or argument as to why it is not an acceptable alternative. It is simply dismissed—in essentially the same way it has been dismissed for the past 50 years.

As the Commission report states very clearly, and just as its Chairman Sandy Peel has rightly continued to insist, the forest economy of British Columbia is facing a desperate future. The report also makes it clear that fundamental, sweeping changes are needed. In this context it is irresponsible to reject without discussion the option of transferring ownership of some portion of the province's commercial forest land into the private sector.

One of the purposes of writing this is to present some arguments in favour of diversifying the ownership of BC's forest lands, and to suggest a mechanism for accomplishing this. There is, I suspect, an unstated reason behind the continued opposition to private forest ownership, not only in BC, but in most of Canada. That reason has to do with the distrust many bureaucrats, resource professionals and others feel toward the ordinary citizens of the province, particularly those who work in the woods.

Aside from noting the existence of the attitude, I do not propose to delve into the psychology of the centrist resource technocrat. What I would like to do is encourage a thorough public debate on this fundamental policy issue. The manipulation of public opinion that was initiated with Orchard's memo and which, ever since, has been a feature of resource politics in BC, has played a part in leading us to the edge of the economic and ecological precipice where we sit today. The continued rejection in BC of a globally successful principle of forest tenure without first holding a free and open public discussion on the subject is folly.

Forest Ownership

Today in BC the province owns 95 percent of the commercial forest land. One percent—most of it Indian reserves—is said to be under the control of the federal government. Another one percent is held by small, private owners. And the remaining three percent is owned by forest companies. The unique nature of forest ownership in BC is apparent when it is compared with that of other jurisdictions (see Table 1).

Table 1

INTERNATIONAL FOREST LAND OWNERSHIP
(By percent)

Country	Public	Private	Corporate	Other
Norway	18	75	-	6
Finland	23.6	65.3	7.4	3.7
Sweden	26	49	25	-
France	12	70	-	18
U.K.	50	50	-	-
W. Germany	31	44	25	-
Yugoslavia	40	60	-	-
USSR	90	-	-	10
Japan	32	57	-	11
US	28	59	13	-
Canada	92	(8)		-
BC	96	1	3	-

It is clear from these figures that Canada as a whole, and BC in particular, share membership in an exclusive club with the Soviet Union—which, of course, does not exist any more. It would probably be instructive for BC policy makers to examine and compare the Soviet system with our own and consider the implications of its collapse in the light of conditions developing here. The other point worth noting from the above figures is that the private sector, which is the largest ownership group in every country outside Canada and the USSR, is composed of many small owners, the bulk of them with holdings of less than 100 hectares.

The History of Tenure

The story of how BC acquired this structure of land ownership begins in the colonial era, before any legislation existed governing the issue. When Captain Edward Stamp made application to Governor James Douglas for timber to supply a mill at Port Alberni in 1859, he was granted title to 17,000 acres of land. This land-granting practice continued until 1896, when legislation was passed prohibiting the sale of timber land. The laws were not rigidly enforced until 1912.

In 1865, however, a Land Ordinance made it possible for the Crown to grant timber harvesting rights on public land, without relinquishing title to the land. The forest industry's prime concern was timber; whether the land came with it was irrelevant, as long as the various forms of leases and licences which became available provided access to the necessary timber. After these leases became available, the only significant transfers of forest land ownership were to promoters as an inducement to construct railways. The largest and most valuable of these was the 1883 grant of 1.9 million acres on the east coast of Vancouver Island for construction of the Esquimalt and Nanaimo Railway. Large blocks of the E & N lands were sold to lumber companies in later years and constitute most of the corporate and private forest lands in BC today.

The principle of granting rights on timber only was retained until 1947, when Orchard's recommendations were enshrined in law. The Forest Act created a new kind of tenure, Forest Management Licences, which consisted of leases on public land. Provisions were also made for Farm Woodlot Licences, but few were ever granted. When Forest Management Licences—later changed to Tree Farm Licences

(TFLs)—were first issued, they were for perpetual terms. At a later date, 21-year TFLs were granted, and in 1978 the term of all were set at 25 years with virtually automatic replacement provisions.

Orchard's original idea, he explained in his memoirs, was that a few hundred of the companies then logging in the province would obtain TFLs and transform themselves into land-based forest managers, harvesting a perpetually sustained yield of timber. Unfortunately, Orchard had overlooked some factors and failed to foresee others. He did not see that the granting of these licences, at no cost, constituted the bestowal of an immense windfall capital gain on the recipients. The acquisition of future cutting rights inherent in the TFLs was worth a lot of money and cost practically nothing.

Orchard did not foresee that, following the Second World War, vast pools of capital would be available, and that it would be used to take over some of the established forest companies for the primary purpose of cashing in on the easy money to be made by merely obtaining a TFL. And, finally, he did not appreciate the venality of some politicians—or, if he did, he chose to ignore it when the Liberal-led Coalition and the successor Social Credit cabinet ministers began accepting bribes to issue TFLs.

Most of these first TFLs were very large—the first in Forest Minister Edward Kenney's riding was for 2.3 million acres. That few, if any, of the recipients saw them as an opportunity to do anything more than harvest timber is underscored by the lack of protest when their terms were reduced in later years.

Until recently, one of the more enduring myths of provincial forest administration was that the TFLs were the best managed lands in BC. As we have seen even Pearse, one of the most astute forest analysts, repeated this assertion in his Royal Commission report. It is only in the past few years, with disclosures of deliberate mismanagement on TFL #1, doctored waste assessment reports in the Queen Charlotte Islands and substantial overcutting of South Island TFLs that led to mill closures, that management deficiencies have been recognized, although not universally acknowledged.

The Rationale Behind the Tenure System
Since at least 1918, when the argument was first advanced in a federal Commission of Conservation study, a rationale has evolved

to defend continued state ownership of forest land:

> The advantages of this distinctly Canadian system were easily recog-
> nized, and, as a result, the province has retained an interest in and a
> control over by far the greater part of its forest resources; at the same
> time it has supplied the lumber industry with abundant raw material
> at reasonable prices.

This basic argument—that after careful and wise consideration,
the province's late 19th-century legislators chose to retain public ownership of the forest lands for reasons of public interest and to insure
the forests would be well cared for—has become one of the most
enduring myths of public policy. Only once has it ever been substantially challenged.

Advantages of Private Ownership

On February 26, 1947, Fred Mulholland, the chief forester of the
Canadian Western Lumber Company, gave a speech to the Canadian
Society of Forest Engineers in Vancouver titled "The Relation Between State and Private Forestry." Until a short time before it was
delivered, Mulholland had worked in the BC Forest Service, directly
under Orchard, and was one of the most respected foresters in Canada. For years he had been a persistent and articulate advocate of
increased private forest ownership in BC, and, in 1939, he had presented a cogent argument for his position before the Society.

Mulholland was far from being alone in his thinking. At about the
time Orchard was penning his memo, at least two forest companies
operating on privately owned lands, Bloedel Stewart & Welch and H.
R. MacMillan Export, had launched silvicultural programs and were
embarking on the type of management systems followed by
Weyerhaeuser in the US and other corporations in Europe.

When it became clear the Orchard plan of continued state ownership was to prevail, Mulholland left the Forest Service—in fact, he was
probably forced out by Orchard, who in a 1946 speech to the Society
in Regina had said, "There seems to be no good reason why exclusive
Government ownership and management of forest lands should not be
successful and satisfactory."

This was clearly an important topic, since Mulholland was asked back to address the Society a year later, where he raised the issue of the origins of the prevalent policy:

> It is sometimes said that this condition [of exclusive state ownership] indicates that our Governments have followed almost from the beginning a wise policy of State ownership in order to conserve the forests for perpetual production. I have no criticism of the distinguished gentlemen who have expressed this opinion; a few years ago I almost believed it myself. But a more careful study of the actions of the Legislature in disposing of land in the earlier days of the Province does not support this view.

Mulholland then went on to demonstrate that the government's purpose in prohibiting further transfers of land ownership had nothing to do with reforestation or a desire to insure the forests were well managed. The real intent was to "get a greater revenue for its timber from the royalty or deferred payment on leases and licences." In other words, the intent of the government in abolishing Crown grants of land was to force up the royalty payments due to the government on the timber already available through leases or licences.

Because the debate that occurred in the 1940s is the one and only time the subject has ever been seriously discussed—albeit within the restricted confines of the Society—it is worth examining in more detail. In his 1939 speech, Mulholland had used the example of other nations' success in private forestry. In his 1946 speech Orchard, after defending the proposition of exclusive Government ownership, conceded:

> It is my own opinion that Government is best able to own and manage forest lands for timber production, but that a reasonable measure of private ownership and management ranging up to, but not exceeding 50 percent, and preferably less than 50 percent, would introduce a healthy competition and act as a desirable tonic on management policy and progress.

A year later Mulholland responded with an appeal to some fundamental democratic principles:

It is certainly the Government's right and duty to see that the private individual does not use or misuse his property to the detriment of his fellow citizens, but for the Government to refuse to allow private citizens to acquire land for proper use, is to admit that it has a low opinion of its people and mistrusts its own power to control abuses. If the control exercised by the force of public opinion and custom is not adequate surely it can be achieved by education or, if necessary, by law and regulation.

He went on to suggest:

Reduced to its simplest terms forestry is, after all, the business of growing trees; it is a form of agriculture. It is the use of a natural resource for human sustenance and enjoyment. In a democracy of reasonable intelligence it is not really necessary for this occupation to be reserved for men hired by the Government, nor even for tenants of the Crown. Independent citizens can undertake it quite satisfactorily, just as they use the natural resources for farming, mining and building homes and gardens. Forests are grown that way in the older countries where forestry is far more advanced than it is in Canada.

Mulholland ended his argument:

Until the Government has its own tremendous forest estate in satisfactory condition it should welcome those who wish to relieve it of some of its burden, especially as private land will contribute to the Treasury through taxation while the trees are growing (with Government ownership no revenue would be obtained).

Mulholland was by no means the only proponent of increased private forest land in Canada. The 1945 annual meeting of the Society was presented with the same conclusion, following a different argument, by A.J. Auden, of the Abitibi Power and Paper Company of Port Arthur. He argued a case based on social objectives:

Forest management is not an end, but a means to an end. The real goal is public welfare. Public welfare always demands social values first. As social values are created, economic benefits for all concerned follow as a matter of course . . . if we set as our ultimate goal the establishment of a resident population in the forest, we will do far more for the cause of forestry, and do it more quickly, than if we set forest management as our goal, and bring in people just to practice forest management.

Auden articulated a vision almost 50 years ago which is today more relevant than ever:

We have spent these past 250 years or so on this continent in restless movement, recklessly skimming off the cream of superabundant resources, but we have not used the land in the true sense of the word, nor have we done ourselves much permanent good. It's high time that we enlarged the family and settled down, not for a hundred years, but for a thousand, forever . . . everything points to thousands, if not a million or so, of small forest holdings, individually and intensively managed, passing from father to son, from son to grandson and so on.

Auden went on to outline the creation of a form of forest tenure on which

freehold right to forest homesteads should be obtainable after a period of probation on a "use-value" basis . . . I think there is a very strong case to be made out for a gradual transfer, under careful but constantly lessened control, from public to private ownership—not large chunks of forest land becoming outright corporation assets, but small, family-sized [holdings] . . .

Ten years later, during cross-examination before the second Sloan Commission, in 1955, Orchard made his last recorded statement on the issue of private ownership:

I think that at our present stage of development one-half of our lands in private ownership would be a wise step. I don't think we can do it. I don't think we ever will. We've got ourselves involved; we've sold the idea of Government ownership so thoroughly, we have such a strong minority element of socialism, I don't think the people would ever let us sell the land. But all those things aside, if we were just looking for the greatest benefit for the most people over the longest period of time, I think we should dispose of one-half of our land in private ownership, and the rest of it, if it were possible, into forest management licences.

Sloan, not unlike Peel 35 years later, ignored completely the option of moving more state land into the private sector. Then, as now, it was as if the words had never been spoken, the question never raised.

But Sloan, like Orchard and other optimists of that era when a surplus timber supply still existed, could comfort himself with the possibility that things would work out for the best. Today we do not have that comfort. We are stuck with a system of forest administration that does not work. Not only have we ended up with a forest industry that often is destructive to the environment in which it operates, we also have an industry in a state of economic collapse, with thousands of people losing their jobs and the future of numerous communities in the province threatened by the failure of their local economies. The ultimate failure of the system is that in the 50 years it has been in existence, it has not built into itself the silvicultural capacity needed to sustain it.

The Concentration of Corporate Control
One measure of this failure is the continuing concentration of control over timber harvesting rights. In his Royal Commission report, Pearse called the tendency toward fewer large companies controlling a larger amount of the action a matter of "urgent public concern." Since he published his report, the 20 largest forest companies have increased their control of harvesting rights from 74 to 86 percent. The share of cutting rights held by the ten biggest companies has increased from 37 percent in 1954, to 59 percent in 1975, to 69 percent in 1990.

One of the lesser consequences of this corporate concentration is that many of the original recipients of TFLs have withdrawn from the province, their briefcases stuffed with the windfall capital gains they received for the sale of cutting rights for which they paid nothing. Not only do the successor licensees have to bear the costs of this payout, but they claim, with justification, that they paid for these rights and if the province wants to cancel them it will have to pay compensation. This condition rewards the quitters, and provides a substantial financial disincentive to a provincial government contemplating a change in the tenure system.

Not only have the corporations become bigger, but they have changed in character. In 1954, most of them were simply big forest companies; by 1975, they were even bigger multinational forest companies, with integrated forest product operations around the world; today the owners of many of these corporations are mere investment companies—money managers in distant high rise towers who, in the words of H.R. MacMillan, the founder of the province's largest forest company, "never having had rain in their lunch buckets would abuse the forests."

For a time, the strategy of some of these corporations—Fletcher Challenge and MacMillan Bloedel being the most obvious—has been to transform themselves into capital intensive pulp producers. To the bright young lawyers and MBAs in Toronto, Montreal or Wellington who direct the directors of these companies, this is a very sensible strategy. Basically, it amounts to this: dispose of the labour intensive components of the company, namely plywood plants, sawmills and any other solid-wood mills that employ a lot of people. Retain and modernize the capital intensive pulp mills that employ few people, that pay for themselves in three or four years at the top end of a good market cycle, and that can be shut down in poor economic times. In the woods, go for low-rotation, plantation forests that can be harvested periodically by a small labour force using million-dollar feller bunchers, grapple yarders and other costly machines.

If there is anyone in this province who doubts where this scenario will lead us they should take a drive down the Olympic Peninsula. What they will see, travelling south from Port Angeles, is an unbroken mosaic of "intensively managed" forests. If they get close to a logging operation they will see a very few people operating very expensive machines, clear cutting second-growth forests that are no more than 60 years old. On the highway they will encounter an unending stream of trucks hauling logs, few of which are more than 18 or 20 inches in diameter.

At the end of this highway they will come to Grays Harbor, and the ancient—by Pacific northwestern terms—lumber towns of Aberdeen and Hoquiam. These communities sit in the centre of one of the most productive ecosystems on the face of the earth. They were among the earliest centres of industrial activity in this part of the continent, and until recently had a diversified array of primary and secondary forest products plants.

Today the Grays Harbor area is one of the most depressed in the northwest, with 35 percent unemployment rates. The elegant old brick and stone commercial buildings are run down and decaying, the new mall is half closed, plywood boarded over the windows. Most of the mills that continue to function are the pulp mills, utilizing the young second-growth logs from the peninsula. The only people making money here are the corporations who own the pulp mills (at least when prices are up) and the financial institutions that loan money to buy the million-dollar logging machines methodically working their way back and forth across the Olympic Peninsula.

This is what the corporate dominated forest industry has done to itself; the dislocation caused by restrictions on logging to protect northern spotted owls will come on top of these self-inflicted wounds, but, curiously, the bitter and resentful unemployed US forest workers have been persuaded to direct their anger at the owls and those they perceive to be their protectors.

This is the future of British Columbia, if we continue on our present path. Aberdeen and Hoquiam are no different from Port Alberni, Duncan, Powell River or Campbell River, 20 or 30 years into the future. What got Aberdeen to its present state was being surrounded by lands owned or controlled by a few very large forest corporations or the US government. This lack of diversity and competition in the woods helped create a certain type of capital intensive processing industry that does not need as many workers as the previous, more diversified, economy did. In BC, we have forest-dependent communities surrounded by even larger, more monopolistic, tenure holders. In some areas, particularly the South Island, the solid-wood mills are already closing because of timber shortages.

The irony and the tragedy of the situation, on the Olympic Peninsula as well as in the BC communities mentioned above, is that they all sit in the midst of lush, young forests growing on some of the most productive forest land on the face of the earth. There are hundreds of thousands of hectares of second-growth forest that would benefit enormously from silvicultural treatments, as well as providing substantial volumes of timber from thinnings and other types of improvement cuttings. But there is practically no one employed in this task. All of these communities should be centres of wealth and vitality. Instead, they are all in a state of economic decline and their citizens are reduced to squabbling among themselves over the remaining scraps of old-growth forests.

The Tenure System's Flaws

In BC there are two basic weaknesses in the existing tenure system. The first problem is a lack of tenure security. Regardless of the size of the licence area—Tree Farm Licence or Woodlot Licence—the land-based tenures are for limited periods of time, subject to a constantly changing array of rules and regulations and at the mercy of changing political fortunes. Who can say what would have happened if the original TFLs had retained their perpetual terms? It is now academic. What is clear is that long ago most TFL licensees determined that the state owners of the land they leased could not be counted on to provide the kind of continuity and security required to manage forests. So they adopted another, shorter term strategy—maximize short-term revenues, then sell the whole show and clear out. Instead of treating them as management tenures, the licensees used them as harvesting tenures where silviculture was an obligation instead of an economic opportunity.

The situation is no different on Woodlot Licences. In spite of their best intentions there is not a single Woodlot licensee who is not in the same boat as the corporations holding TFLs. None know from one year to the next, one government to another, what will be the economic consequences of their decisions. The entire system provides a built-in incentive to forego long-term gains in favour of immediate returns.

It is easy for people at a distance, whose personal, family income is not on the line, to say that licensees, Woodlot or Tree Farm, should make this or that kind of silviculturally sensible choice. It is another matter altogether for those holding these licences to see their decisions undermined by the rulings of bureaucrats following the impeccable logic of their ministries. In this kind of situation the best of forest managers will hedge their bets, and take their profits when they can.

The second problem is that a central concept of the TFL system—that licensees were "renting" the land on which they were growing the next forest crop—went by the wayside very early on in the game. For several years, until settlement of the 1986 softwood lumber tariff battle, it was argued successfully that, since the province was the landlord, it was the province's responsibility to pay for looking after the forests. This was a wonderful system for the licensees, who had their silvicultural costs paid for, and for the army of bureaucrats in Ottawa and Victoria who dreamed up and administered the forest management programs. Needless to say this did not work: after a few years the governments got tired of spending unending amounts of money on this sort of thing, and many of the practices were silviculturally suspect—an issue that takes us to the heart of this second major flaw in the tenure system.

The system is based on the premise that silviculture is something that can be administered from a distance, that it is a mechanistic activity that can be planned by scientifically trained experts, administered by centralized bureaucracies and carried out by unskilled, uneducated and unmotivated workers. What has been created is a centralized, monolithic intensive forest management network of bureaucrats, contractors and seasonal forestry workers. Its character, organization and effectiveness is about on a par with the centrally controlled agricultural system that failed so miserably in the former Soviet Union. It is simply not possible to decide from an office in Victoria how to manage every hectare of forest land in BC, yet that is essentially how the system works.

The consequences of the existing system are increasingly apparent. The supply of timber provided by the system is shrinking and leading to mill closures. Jobs are disappearing, and the economic stability of timber-dependent communities is threatened. Instead of transforming the industry and providing opportunities for displaced workers to manage the young forests, the inflexible industrial structure is simply made smaller and those thrown out of work have to fend for themselves in another industrial sector.

The poor management practices of the past are also affecting forest values other than timber—fisheries, water resources, wildlife, and so on. The deteriorating economic situation is used as an argument to allow destructive practices to continue, for the sake of jobs and community survival. The willingness in the past of management, union and government leaders to allow these practices, and to argue for their continuation, has led to the emergence of a hostile public, unwilling to let the industry continue on its present course. For the first time in its history the province's major industry has lost the confidence of the province's people.

And, finally, it must be recognized that the expenditure, over the past 15 years, of hundreds of millions of public dollars on misconceived forest management programs has failed to create a forest management capability. We are not optimizing timber values, we are still unable to manage effectively for non-timber values and we are foregoing tens of thousands of permanent jobs, a future supply of high-quality timber and the potential to stabilize hinterland economies.

The forest industry has evolved into a form of organization that could be called monopoly corporatism. As these corporations insist, they compete among themselves in the global marketplace, but from the point of view of the communities in which they operate, they are monopolies—they have eliminated competition from the logging business, they do not sell their timber on an open market where it would be available to ever more efficient mills, and they are the major threat to the remaining vestiges of a resident business community. On a provincial level, they, their unions and the Ministry of Forests constitute a single, monolithic entity—besieged by a hostile public and united in a perceived necessity to resist change.

The Case for Diversification

In the face of this deteriorating situation it is easy to make a good case for a more diversified forest ownership. The oldest and simplest argument is that the diverse forest ownership of other countries appears to be a critical factor in the stability and vitality of their forest economies. After the collapse of the only other jurisdiction with an ownership structure similar to ours, it would seem prudent, at least, to take a good look at the situation in countries with populations of resident forest owners to determine how their experiences might apply in BC.

A major reason for considering abandonment of the policy of exclusive public forest ownership is that 50 years of practice has demonstrated fairly conclusively that the prevailing tenure system does not provide the kind of long-term security required. There is also a question of social values. Managing our forests at a level that would maintain the economic benefits we have enjoyed in the past means tens of thousands of people eventually will be engaged in this work. Do we want a society in which a large, landless labour force is employed by a small number of big, remotely-owned corporations under the direction of a huge and expensive state bureaucracy? The alternative is that we opt for a more diversified socio-economic structure that includes a significant component of autonomous, independent, privately owned forest farms. In agricultural terms we can choose between the models of the state-run, communal farms that failed in the Soviet Union, or the North American family farm that was the basis of the most productive agricultural system in the world.

Peel Commission Recommendations

In this regard, the recommendations of the 1991 Forest Resources Commission report are worth examining. The FRC's basic proposal was, that after determining the commercial forest land base, to turn that land over to a Forest Resources Corporation (Forestco), a Crown corporation much like BC Hydro. Forestco would then re-issue, essentially, the existing tenure agreements, replacing Tree Farm Licences with Resource Management Agreements and some Forest Licences with Wood Supply Agreements. The TFL holders would keep what they have now, while existing Forest Licence holders would lose 50 percent of their cutting rights. This volume would be allocated, in unspecified proportions, to "small area-based tenures managed by communities, Native Bands and woodlot operators, etc." Forestco would itself manage "sufficient forest lands from which wood will be available on a volume basis in order to retain forest management flexibility."

There were many more details in the FRC proposal, but, with one exception, none of them would have significantly changed things from the way they function now. Natives would have received some land, more Woodlot Licences would have been created, along with ill-defined "community" forests. There was no mention of increased timber rights for the independent loggers who now work under contract to the big companies and whose arguments in support of small, locally-owned businesses constitute one of the major forces for change.

The one exception was hugely significant: Forestco would have been empowered "to borrow against its asset base—represented by the value of the standing trees, not the land—in order to maintain continuous stewardship in times of economic down-turn." An appended consultant's report estimated Forestco would have been able to borrow between $8.3 and $8.5 billion.

Judging by the lack of enthusiasm on the part of either the past or present governments, and the reduced status of the Commission, the Forestco solution does not appear to have wide support and the government may be considering another inquiry of some sort to re-examine various policy issues, including tenure.

It is difficult to see how the insertion of Forestco into the existing tenure structure would have created the kind of confidence required for private companies or individuals to get into the business of managing forests. The assumption that a Crown corporation would provide any more security than a government department is neither demonstrable nor credible. It would simply add another level of bureaucracy.

This, however, is not the real danger represented by a concept like Forestco. That lies in the proposed ability to borrow the huge amounts of money indicated by its estimates and then dispense those funds to the companies leasing its forest land. Not only would it mortgage the future of the province's forests, but the government corporation would be in a position to use the money to wield enormous power over the entire forest industry. In any case, there is a flawed assumption built into this proposal that the basic management problem is a financial one and, if enough money is spent on forest management, forest yields will increase proportionately.

On the other hand, what are the alternatives?

Proposals for Changing the Tenure System

Of the numerous proposals that have surfaced recently, probably the most valuable is that of UBC forester Peter Sanders. He suggested a reconfiguration of the industry, so that one-third of the commercial forest land is held by small, private land owners, another third by corporate owners, with the final third remaining in public ownership.

So, how would title to the land be transferred? The obvious method—for the province to sell it to private interests—may not necessarily be the most advantageous. In the first place, bare land values on large tracts of forest land have traditionally been quite low, even with the small amounts of such land available. If the amounts of land under consideration were placed on the market, bare land prices would drop to insignificant levels and purchasers would only be willing to pay for the value of the timber growing on the land. The province would realize very little revenue from this approach and find it difficult to discourage or bar buyers with no intention to manage the lands for timber production. Additionally, this procedure would tie up purchasers' capital that could be put to better use if it were invested in silviculture and equipment.

The government could use the instrument of title or ownership in a much more effective manner by offering it as a reward or incentive for practising good forest management during a "proving-up" period. This approach involves essentially the same principles inherent in the homesteading laws that were used with great success in developing North American agriculture. By proving one is a good manager, over a term of 10 or 15 years, one earns title to the land. Ownership, under this method, is a reward for good forest management and not simply an option of the wealthy. Any windfall capital gains involved in such a system could be captured by the province if the land was sold; otherwise, public revenues would be taken in taxes on land and income.

If the principle of earned title were applied, the transition from the present leasing system to one of land ownership would be relatively simple and might avoid disruption of the established forest economy. In the case of small holdings, a schedule could be established during which second-growth forest lands capable of supporting a forest manager would be allocated under a modified Woodlot Licence program. A clause could be added to these licences granting title to the land once the licensee has proved his or her capability as a manager for a set term.

The same principle of earned ownership might be applied in the corporate forest sector. The existing TFLs and some Forest Licences could be transferred to new corporations owned by the managers, contractors and employees presently working on these lands. Some of the TFLs may need to be broken down into two or more smaller units to localize their operations. These new corporations would also earn title after proving management capabilities over a fixed period of time.

They could be required to enter into wood supply contracts with the existing corporate licensees in order to provide an orderly transition to a market system of timber distribution.

The commercial forests remaining under public management would be the multiple-use forests having high non-timber values—water, fisheries, wildlife, recreation, and so on. They would still be classified as commercial forests, but under heavier constraints and free from the obligation to support management functions entirely from timber revenues. These lands would include a much larger number of municipally owned or managed forests, such as now exist at North Cowichan and Mission. In some cases it might make sense for Regional Districts to manage forests within their jurisdictions, as is now the case in the Greater Vancouver Regional District's watershed forests. There is just as much need for diversity within the public forest sector as elsewhere, and management by the local governments of Nelson, Port Hardy or Dawson Creek will be more responsive to local needs than if it is undertaken by the provincial Forest Service.

There are undoubtedly many other mechanisms for diversifying the ownership of forest lands. The approach outlined is merely one method that attempts to accomplish the desired objectives without causing enormous dislocation in the established industry, and without requiring the expenditure of vast sums of public funds. In itself such a diversification would bring about an enormous change in the long-term outlook of the industry, as well as providing an entirely new timber supply to the mills from second-growth forests. But it would require some additional changes in the province's system of administering forests.

The first need would be for legislation defining a basic Forest Law that says, simply and briefly, "Thou shalt not screw up the forest." It needs to apply to both state-owned and private land. Virtually every forest nation in the world, except Canada and the US, has such laws that recognize the forests are of such overwhelming importance that ownership of them does not confer the right to abuse or destroy them.

A second need is for a vastly expanded silvicultural research capability. Instead of leaving this task to the federal government—BC's historic tendency—we should assume the initiative in this area and integrate our research activities with neighbouring US states that have the same types of forests. There is also a need for a research and development program to support the creation of a manufacturing sector to produce the technology and hardware required to intensively manage the forests of the northwest.

With the focus of the industry shifted from harvesting to managing there will be great need for a silvicultural education system throughout the province. The theory and practice of forest management should be introduced into the public schools, and the community college network should be mobilized to provide a full array of teaching and training programs to forest managers. A forest extension service that links the research and education components with the forest managers is also needed.

•

The scenario defined above is not intended to be comprehensive. Its purpose is to illustrate that it is desirable and possible to diversify forest ownership in BC. What is required is a full-scale public discussion of the option.

Some might argue that the current New Democratic Party government is an unlikely agency of such a diversification program, due to its historic ideological opposition to private property. Quite clearly this government still has to work out its position on the relative merits of public and private ownership, but given the shift in thinking among left-leaning politicians around the world it is unlikely that BC politicians of this persuasion will continue their past tendencies to view private ownership as an unmitigated evil.

An NDP government is probably the only government the public would allow to transfer public lands into the private sector. No one in their right minds would have trusted either the Bennett or Vander Zalm governments to undertake such a task, but if the current government decided to do so it would probably be able to accomplish the task with the confidence of the people.

Finally, it is important to remember that the objective is not the privatization of the province's forest lands. Diversifying forest ownership is only a means to realizing the social values of the people of BC.

What kind of a society do we want to build in this province? If, as I suspect, the people of BC want a free and diversified society, based upon a sophisticated resource management economy that is in tune with its natural environment, then it makes sense to carefully and openly examine the option of diversifying the basic building block of social and economic organization—land ownership.

Sources

Information on early forest policies was obtained from H.N Whitford and Roland Craig, *Forests of British Columbia*, Ottawa, 1918; and Appendix B in *Crown Charges for Early Timber Rights*, Task Force on Crown Timber Disposal, Victoria 1974.

Material relating to the debate on private versus public land ownership were obtained from various issues of *Forestry Chronicle*, in the years indicated in the text.

Additional material was drawn from the two Royal Commission reports of Gordon Sloan in 1945 and 1956, Peter Pearse's 1976 Royal Commission report, and the 1991 report of the Forest Resources Commission.

For additional information also see David Haley, "The Forest Tenure System as a Constraint on Efficient Timber Management," in *Canadian Public Policy*, 1985, p. 315.

Public Participation

Changing the Way
We Make Forest Decisions

by Bob Nixon

IN TODAY'S TRANSITIONAL WORLD, we have outgrown many of our institutions. Our representative parliamentary democratic tradition has not kept pace with changing public values. Across North America—not just in Canada or British Columbia—public officials find themselves unable to make resource decisions that reduce political tensions. Time and again, governments create public consultation processes, yet most fail dismally. Two commonly used phrases which, through constant misuse, have become meaningless in the context of attempts to solve forest land use problems are "public input" and "public consultation." Government and the forest industry ask us to lend *their* decision-making processes credibility through our participation. Yet the results tend to generate even more discord and public dissatisfaction.

As timber depletion accelerates, residents of logging towns and villages—who once believed their livelihoods were secured by promises of perpetual sustained yield timber management—face the immediate choice of either watching their communities become ghost towns or creating a replacement for the old forest economy. As if this were not enough, the depletion of the province's former economic mainstay has led us to the doorstep of two even more polarizing demands: faced with increasing timber shortages, we tend to plan our cutting with even less regard to the future, and recognizing scarcity of wilderness values elsewhere, the world now focuses attention upon our province with the intent of preserving much of our remnant old-growth forests.

As the problems multiply, many British Columbians desire a new vision of forest use—a land ethic that embraces the new values of a society growing accustomed to thinking "green." The old path has led us to roadblocks, international boycotts, and litigation. We have worn a path toward an outcome known as "zero-sum." Nobody wins.

The impasse is hardly limited to environmental concerns. The Forest Resources Commission (FRC) of British Columbia—after nearly three years of investigation—warned us of imminent economic collapse in the forest sector. Our current path, FRC Chairman Sandy Peel said, is taking us toward a 50 per cent contraction in the size of our forest industry, along with the loss of tens of thousands of forest jobs and provincial revenues. All this will occur, it seems, even if we say no to any further protection of old-growth forests, and to integrated forest management that protects all forest values, and not just the short-term timber supply.

We seem to be stuck with an obsolete system. Increasingly, official government documents tell us that our approach over the past several decades has not delivered satisfactory forest management. In early 1992, for example, the Ministry of Environment's habitat branch asked a Nanaimo fisheries consultant to conduct a sampling audit of logging impacts upon Vancouver Island streams. The results proved a devastating indictment of current forestry practices. Indeed, even some senior Ministry of Forests officials seemed truly shocked by the evidence. How could this be? Both the forest industry and the Ministry of Forests have been telling the public that forest practices problems were all "mistakes of the past." "We don't do that any more." But the reality of forest management does not match the message. Unable to hide the problems, both industry and the Ministry of Forests seem paralyzed, incapable of identifying a means to change their ways.

Some deputy ministers and other senior bureaucrats will candidly admit to this impasse, at least in private. On the other hand, they also seem powerless to effect change. A great number of our over 3,000 Ministry of Forests personnel are intensely concerned about the inability of their employer to provide reasonable stewardship for all forest values. Senior forest industry executives also acknowledge the need to find the means by which our always simmering—more often boiling—forest disputes can be resolved. All interests, it seems, recognize the need for change.

Most of the answers to our difficulties in managing forests are not to be found by conducting more research or in the exercise of traditionally inept "leadership" by politicians. The solutions have more to do with rediscovery of our basic democratic rights. This involves envisioning a new "community" approach to problem solving. Asking politicians to make "tough decisions" to settle land use disputes is not the answer, although tough decisions will have to be made. Tough decisions, made in the absence of democratic consensus building, only provide temporary, short-term solutions. The underlying conflict becomes worse. The fundamental problem—the reason why forest disputes grow larger with each passing day—is the absence of democratically fair public participation processes. Current forest planning processes systematically exclude significant sectors of society with legitimate interests in forests. If our democracy is to be functional in the realm of forest use, then all those affected by forest use decisions must be afforded a reasonable opportunity to participate directly in forest use decision-making.

The Ministry of Forests is intended to administer forest resources fairly on behalf of all private citizens of British Columbia. The tension between the rights of citizens and the prerogatives of the State are as old as democracy itself. Nevertheless, as protector of the rights of private citizens, it is the duty of public administrators (a Ministry of Forests) to ensure the rights of citizens are respected—and protected from the excesses of the State. In a democracy, there can be no other role. Yet senior Ministry officials now confirm their inability to fulfill this responsibility. They admit, privately, to previous political instructions under successive Social Credit governments to serve only the major forest industry corporations and either ignore or exclude all others, including the interests of the provincial Ministry of Environment, and even the federal Department of Fisheries and Oceans which has jurisdiction over salmon habitat. Having boldly declared the reality of past public administrative abuse, officials seem almost eager to acknowledge the need to bring the private citizen, non-government organizations, and other government agencies into a new decision making structure which respects the rights of all with an interest in the forests, and to treat each interest equitably and fairly. How then, do we accomplish this?

The way to break the impasse is to change the way we make forest decisions—by inviting all interests to share responsibility in inventing new processes and envisioning options that will, over time, build consensus. This is called "shared decision-making."

The Present System

In our parliamentary democracy, we rely upon individual Cabinet ministers to make decisions on our behalf. Graham Bruce, minister of municipal affairs in the former Vander Zalm government, once wrote that ours is an "arrested parliamentary democracy." He meant that even when compared to the "Mother of Parliaments" in Great Britain, our own decision-making processes have not evolved to keep pace with changing times. The traditional approach of most elected officials is: "You've elected me, now get out of my way and let *me* govern. If you don't like what I'm doing, then don't vote for me when we go to the polls in five years." In today's fast changing world, such a philosophy of governance is out of step with public expectation. In musing on the outcome of the Charlottetown Constitutional Accord Referendum, Prime Minister Mulroney said:

> I always thought, quite frankly, that under the British parliamentary system, a referendum was a kind of abdication of responsibility. I've changed my mind over the years. I've come to recognize that in a modern, pluralistic society like ours, people do indeed require a much greater degree of participation than a kick at the can every four years. (*Globe and Mail*, October 31, 1992, p. 1).

Moreover, the Ministry of Forests has enormous discretionary powers, and neither the public nor Members of the Legislative Assembly have much knowledge of how they are exercised. Elections every four to five years are not fought on the basis of forest overcutting or destruction of marbled murrelet nesting habitat in old-growth forests.

Graham Bruce also served as chairman of the British Columbia Legislature's Select Standing Committee on Forests and Lands. Prior to his chairmanship, this legislative committee had failed to conduct almost any public business during the previous 20 years. In effect, the role of elected members of the legislature had been diminished in favour of executive discretion by individual Cabinet ministers. In the legislature itself, with the exception of cursory examination of forest issues during the ritualized annual forest ministry spending estimates debate, matters concerning the forests and their dependent industries were (and continue to be) dealt with entirely outside the legislature.

Deals are made behind the closed doors of ministers' offices, in the board room of the executive of the Ministry of Forests, and in countless private meetings between Forest Service officials and the Ministry's only legally recognized client, its timber licensees—the logging corporations.

The failure of elected legislators to effectively keep tabs on the public interest is not, of course, restricted solely to forest issues. The loss of credibility for public officials is widespread, and growing. The problem, it seems, cannot be corrected by simply throwing out one government and electing another.

Within the Ministry of Forests itself, the forestry planning and integrated management functions are organized within a split hierarchy designed to restrict interference with the Ministry's historical mandate to liquidate British Columbia's old-growth forests. For example, according to the Ministry's own rules, if a forester within the Integrated Resources Management Branch wished to hold a meeting with a field operations counterpart in one of the 42 Forest Service District Offices scattered throughout British Columbia, a memo to request authorization for the meeting would have to be approved by the provincial Chief Forester, then forwarded to the Assistant Deputy Minister of Operations. The Operations ADM would then have to approve the request of the Chief Forester, forward his endorsement to the appropriate regional office which, in turn, would notify the district office that authority has been granted to allow the district staff person to attend a meeting with the forester from the Ministry's Victoria-based Integrated Resources Branch. The procedure has nothing to do with financial cost control. Rather, this and similar internal procedures are designed to prevent the "show" branches of the Ministry of Forests—Integrated Management and Recreation—from interfering with the goal of providing trees to the forest industry.

Such a rigid internal structure serves to isolate all those within the

Ministry who are alleged to have responsibility for multiple use, forestry planning, and integrated management. The result is to ensure that the Operations Division of the Ministry of Forests has an unfettered hand to exploit the forest without interference. Those Ministry staff charged with the theoretical responsibility to consider all forest values, including those who consider the need to protect the long-term productivity of the forest to even produce timber, are isolated and reduced to the role of occasional participants in Ministry of Forests affairs. The reality is that a "Berlin Wall" exists between the Operations Division of the Ministry of Forests and those charged with forestry planning, integrated management and recreation. One former provincial Chief Forester actually referred to those behind the wall in Operations as "that evil empire."

The Lesson of Clayoquot Sound

A decision made early in the mandate of the newly elected (1991) BC New Democratic Party by its forest minister, Dan Miller, provides an excellent example of a generic problem. In this case, the issue involved the transfer of a portion of a major tree farm licence (No. 46) on Vancouver Island from Fletcher Challenge Canada to Interfor (part of the Sauder Group). The tenure area which Fletcher wished to be rid of included all of its Crown (public) forests within Clayoquot Sound.

It was a smart strategic move from Fletcher's point of view. One glance at a satellite image of southern Vancouver Island shows that much of the remnant coastal temperate old-growth rain forest is concentrated within Clayoquot Sound. Fletcher reduced its exposure to environmental criticism by getting rid of public tenure in this "hotspot."

Only one political-legal problem needed to be overcome, jointly, by Fletcher and Interfor in order to consummate the transfer of mill assets and public forest tenure. Namely, under the Forest Act, public forest tenure cannot technically be bought or sold in British Columbia. Thus, by law, when a company sells a manufacturing plant (or two), as Fletcher did in this instance, the Minister of Forests must authorize the transfer of public forest tenures deemed to be linked to the mills. Fletcher needed to be assured that the Minister would approve the transfer of timber rights, as did Interfor, before final signatures could be affixed to the mill-sale documents.

So, too, did the residents of Clayoquot Sound require assurance that their interests would be respected when the licence area was transferred to a new tenure holder. They didn't get it.

For the previous three years, a provincial government initiative (in which the Ministry of Forests participates) had been in place attempting to resolve the forest land use dispute in Clayoquot Sound, which centers around how to provide protection for remnant old-growth forests while ensuring a continuation of the forest industry. The applicable section of the terms of reference asked participants to consider how to address the need to protect intact wilderness areas with "minimal disruption to the forest industry." The Clayoquot Sound Sustainable Development (Task Force) Steering Committee, with representation from each regional community and some sector interests such as labour and small business, had been attempting to function according to the principles of "negotiated, interest-based, consensus building"— principles which some government ministers do not themselves consider democratically legitimate in a British parliamentary system of government.

Forests Minister Miller did abide by the Forest Act in agreeing to the transfer of timber rights, but totally ignored the people of Clayoquot Sound in making his decision. He may have made the right decision, attaching certain conditions to the transfer, but the lasting impression in the minds of many was of a minister failing, once again, to consult. Or if he did consult, he did so behind closed doors with only those special interests deemed powerful enough to capture his attention. The minister consciously decided he could ignore the *ongoing* efforts of a myriad of citizen groups who had already invested the previous three years in an attempt to build consensus on how to manage the forest resources of Clayoquot Sound to ensure long-term sustainability.

Three weeks prior to the minister's announcement of his decision on the licence transfer, I asked Dan Miller what step he would take to consult with the affected local interests—no response. On the day of the announcement, in an interview with the media, Miller attempted to pass off his failure to consult with affected local interests (Native communities, environmentalists, small business, etc.) as unimportant. Time and again, during the past three years, the provincial government re-affirmed its commitment to the very public Clayoquot Sound

Sustainable Development process, always claiming, when faced with criticism, that the Clayoquot process was the means by which decisions would be made concerning the future use of natural resources within Clayoquot Sound. Yet the Minister of Forests completely disregarded this public consensus-based process in reaching his tenure transfer decision. Forests Minister Dan Miller failed to permit the participants in the Clayoquot process to be a part of a decision which fundamentally affects their deliberations. Miller demonstrated that shared decision-making with the affected public was a concept foreign to his definition of governing in a democracy.

Based on parliamentary tradition, the Minister of Forests was elected to govern. Miller made his decision on the licence transfer according to law. There was no legal requirement to consult with the Clayoquot Sound Sustainable Development Steering Committee. Indeed, the minister might argue—as others have so argued—that to consult with the Steering Committee and thereby seek consensus of all "stakeholders" or community "interests" prior to making his decision would be an infringement upon his democratic right and responsibility to govern.

Jobs for Some, But Not for Others

The original impetus for creating the Clayoquot process came about because the small community of Tofino prepared a report on the urgent need to consider the use of forest resources within Clayoquot Sound in a much broader context utilizing sustainable development principles. Their view was based on the pioneering work of the Brundtland Commission's report, *Our Common Future.* They were encouraged by endorsements for this approach from both the federal and provincial government. Tofino feared that continuation of logging at currents rates and using traditional clearcut harvesting methods meant the eventual liquidation of its unique old-growth forests, with a consequent loss of its thriving tourism industry. As well, residents had become convinced that current logging rates and practices were unsustainable and would, if continued, wipe out future logging jobs and revenues. Later, as we have seen, the Forest Resources Commission Chairman Sandy Peel would confirm that even without any old-growth forest preservation, the forestry industry was doomed to suffer a major economic contraction due largely to overcommitment of the timber supply.

Tofino's report was eventually "adopted-in-principle" by the former Social Credit Cabinet. In fact, there was a race to capture Cabinet's attention. MacMillan Bloedel and Fletcher Challenge rushed to be the first to present their vision for Clayoquot Sound to the provincial Cabinet. At the end of the day, Cabinet responded to the intent of Tofino's community-based sustainable development proposal, but expanded its scope to encompass the broader regional timber supply area, i.e. the mill town of Port Alberni where a portion of the wood to be cut in Clayoquot Sound was destined.

Tofino had called for a community-based analysis of future options. The proposal would have begun at the local level, built community consensus, and then expanded to consider the broader regional economic zone extending to Port Alberni. Eventually, after much delay, a terms-of-reference was prepared by the Premier's office which launched the Clayoquot process, although the original Tofino report seemed to become lost in the bureaucratic shuffle. The process charted a different course, driven by political expediency, to assure a dominant place at the negotiating table for the users of timber who tended to perceive Tofino as "the enemy." As a result, consensus principles were an early casualty. One year into the process, the environmental sector representatives at the table resigned, citing a decision by the provincial government to proceed with logging in selected intact old-growth areas within Clayoquot Sound, and due to a long-simmering, intense dissatisfaction over a number of procedural issues which centered on the belief that the overall process was administratively unfair.

At the Clayoquot Sound Sustainable Development Steering Committee meetings, MacMillan Bloedel and Fletcher Challenge demanded, with the support of the labour union which cuts the trees (International Woodworkers of America), that all attention be focused upon job retention for logging and road building crews. The motto became "no job loss," even though the terms of reference spoke of "minimal disruption to the forest industry." Most participants at the table—with the exception of the District of Tofino which had asked for the process in the first place—were consumed with the need to ensure that efforts to preserve remnant old-growth forests did not result in the loss of a single logging or road building job. Efforts by Tofino to address the longer-term economic stability of the forest industry in the region were ruled out of order. "Consensus" in the process was interpreted to mean that all interests had to agree before a concern could even be brought to the table for consideration. Yet, in the meantime, MacMillan Bloedel (one of the participants at the table) wiped out several hundred forest manufacturing jobs in the region. This concurrent loss of some 400 milling jobs in Port Alberni was not deemed a legitimate subject of concern, even though the issue was voiced. Such overt manipulations of the "consensus process" by the more powerful interests at the table have discredited efforts to reform planning and decision-making processes.

The misuse of the consensus approach had led to the meaning of consensus being redefined such that issues of concern could not even be discussed at the table unless, by consensus, all agreed that the concern was legitimate. It was a convenient way for the logging companies to avoid dealing with substantive sustainable development issues. Both MacMillan Bloedel (Tree Farm Licence 44) and Fletcher Challenge Canada (Tree Farm Licence 46) were participants in the Clayoquot process. Then, in December of 1991 without prior warning or discussion—as the struggle to protect unique rain forest ecosystems while minimizing disruption to the forest industry continued—MacMillan Bloedel unilaterally cut logging employment levels in Clayoquot Sound by some 40 per cent. This in addition to the 400 permanent jobs eliminated earlier. The credibility of the process had, by this point, been thoroughly corrupted. The consensus process was not addressing the issues because neither government nor the forest industry brought substantive issues to the table. One of the major questions continuously on the agenda of the three-year-old Clayoquot process had been how to maintain forest employment opportunity in the short term, while attempting to decide the means by which a sustainable balance of resource use could be agreed upon for the future. Indeed, during the first year, participants considered little else. Time after time, those who sat at the table (MacMillan Bloedel, in particular) made significant decisions affecting the future of Clayoquot Sound, but never considered the Clayoquot process to be either an important or legitimate forum. The process was reduced to being a bystander and observer of major events. Meanwhile, the provincial government strove mightily to maintain the illusion that the Clayoquot process was still the place where long-term land use decisions would be made. Instead, all the important decisions affecting the long-term use of forests in Clayoquot Sound were not even debated among participants. The real decisions continued to be made elsewhere.

The Atlantic Salmon Decision

More recently in Clayoquot Sound, another Ministry, this time the provincial Ministry of Agriculture and Fisheries, unilaterally decided that Kennedy Lake (a large fresh water lake adjacent to Long Beach National Park) would be a good place to introduce and rear controversial Atlantic salmon. Since the Sound is blessed with an abundance of native salmon species, many local residents are totally opposed to the introduction of non-native stock, with the attendant (and very real) risk of exotic diseases which could dramatically affect resident, native salmon species.

Following the very same procedural path previously taken by, first, Forests Minister Dan Miller, in transferring tenure from Fletcher Challenge Canada to Interfor, and, second, MacMillan Bloedel, who ignored the existence of the Clayoquot process when deciding to eliminate hundreds of forest industry jobs within the region, fisheries bureaucrats decided to join the anti-consensus parade. Fisheries officials proceeded independently to identify a lake within the Clayoquot Sound study area as suitable for the introduction of Atlantic salmon. Having reached this internal decision, all that remained (from the fisheries bureaucrats' point of view) was to decide how to process individual permit applications from their clients, the aquaculture industry. When the secret plan was uncovered and local residents had a chance to confront directly the government representatives who had been participating in the Clayoquot Sustainable Development consensus process for nearly three years, these same bureaucrats defended their unilateral right to proceed without consultation with the Clayoquot process based upon authority granted in the Fisheries Acts. A provincial fisheries official had participated as a representative on the Clayoquot process for the past three years, and indicated that he was under no obligation to consult with other participants in the consensus process.

The Clayoquot "consensus" process was a conscious attempt to improve the consultation process by which government attempted to resolve land use disputes. It sought to apply principled negotiation and consensus decision-making to the debate over the future of Clayoquot Sound. Yet, a minister of the Crown, a major licensee, and a senior fisheries bureaucrat (to cite only three instances) failed to abide by the principles of fair process. The words had changed, and the speeches of the politicians now included reference to their commitment to sustainable development and shared decision-making, with all affected interests at the same negotiating table in joint problem solving endeavours. However, the actions of these same politicians, bureaucrats and their traditional clients show that their notion of sustainable development, shared decision-making, and consensus, is nothing more than a smokescreen behind which traditional special interest relationships are maintained, unimpeded by a rising tide of legitimate public interest concerns.

Bureaucratic Resistance to Shared Decision-Making

Many industry and political leaders are openly expressing a desire to put an end to forest confrontation. Having tried everything else in the book—lobbying, unilateral decisions, coercion and bullying—they are faced with the reality that nothing they have attempted recently has reduced political conflict in our forests. Yet there remains one untried alternative—a true consensus, principled negotiation, shared decision-making approach. The main objection from bureaucrats to using forms of principled negotiation and consensus decision-making (the terms are used somewhat interchangeably) is the concern that participating in this type of negotiation may constitute an abdication of personal and legal responsibility. Industry, elected officials and bureaucrats all suffer from the same fear. Their fear of change has paralyzed *our* ability to resolve forest disputes. Or is it fear? Perhaps their objections stem more from an astute awareness that behind the new approach lurks the danger of empowering others to participate in decision-making which affects their interests. Empowerment is indeed at the heart of the new approach.

The shared decision-making approach does have profound implications for both politicians and bureaucrats in the way they conduct the public's business. Their historically unfettered discretionary powers (in the Canadian context) would be dramatically reduced. To some politicians, this, above all else, is something to be feared. To others, often younger politicians, the new approach offers an exciting opportunity to build understanding and consensus in a world where old traditions are, by themselves, no longer adequate to serve the public good. If Prime Minister Mulroney is capable of at least noticing the rising public support for a fairer approach to public decision-making, can provincial officials pretend much longer not to have noticed? Our society is indeed pluralistic, and no longer accepting of politicians who hide behind a stagnant, decades out of date, British parliamentary system, where ministers of the Crown view, with sneering contempt, public participation in decision-making which affects citizen interests—preferring instead to claim that allowing the public more direct participation is, somehow, a kind of dreadful abdication of democratic responsibility.

The argument is a familiar one. In 1976, in Manitoba, upon the introduction of a Cabinet order (a "minute" of a Cabinet meeting) to implement a process of environmental assessment for proposed provincial government projects, deputy ministers and their assistants in both line ministries and agencies offered a similar defense against change. In working for the provincial government in environmental assessment, I assisted in the implementation of the new assessment procedures, and became thoroughly familiar with the bureaucrats' arguments for maintenance of the status quo. The issue, then, as now, was not whether the public interest would be served by this particular change in approach, but rather something more fundamental in human affairs—resistance to change. Regardless of how unworkable old methods may become, we still too often find it difficult to modify our behaviour. Almost any excuse will do.

Real Decisions Are Made Elsewhere

In January of 1992, the provincial Chief Forester John Cuthbert reduced the rate of cut in MacMillan Bloedel's Tree Farm Licence 44, on Vancouver Island, by 14 percent. Cuthbert indicated that there were even more reductions yet to come, as a result of further studies due to be completed within two years.

The Chief Forester's reduction of MacMillan Bloedel's cut was intended to be a first step, designed to reform single-use oriented forest (timber) practices and move the province toward sustainability for all forest values. MacMillan Bloedel immediately launched an appeal against the reduction—and won. It was the first-ever appeal against a rate of cut reduction, launched under provisions of an obscure section of the provincial Forest Act. When the decision on MacMillan Bloedel's appeal was eventually handed down in September 1992, Jasper Stevens, a Tofino resident, fisherman and participant in the Clayoquot process noted:

> MacMillan Bloedel sits as a representative on the Steering Committee which is attempting to balance the protection of the Sound's temperate old-growth rain forests with economic development, while its lawyers claim before a Forest Service appeal hearing in Vancouver that provincial Chief Forester John Cuthbert lacks authority to reduce MacMillan Bloedel's allowable annual cut . . . We all acknowledge, even MacMillan Bloedel's representative in our local process, that this means the cut must be reduced, but outside the process MacMillan Bloedel takes an entirely different position . . . MB must look upon our local sustainable development process as some sort of joke.

The appeal decision also disclosed that, despite Ministry of Forests and industry claims to the contrary, integrated resource management was a new concept in British Columbia, which has never been applied in the calculation of a rate of cut on any Tree Farm Licence in British Columbia. As well, the appeal decision disclosed that even the provincial Chief Forester lacked legal authority to consider the protection of non-timber values unless the tenure licensee (MacMillan Bloedel in this case) volunteered to include consideration of other values in their management and working plan submissions, submitted every five years to the province for approval.

Conflict of Interest?

In early September of 1992, British Columbia's Conflict of Interest Commissioner Ted Hughes presented Premier Harcourt with the findings of his investigation of a complaint against Forests Minister Dan Miller. It concerned the awarding of a timber licence to REPAP, a forest company which owned the mill where Miller was employed in

Prince Rupert, prior to being elected as a member of the Legislative Assembly of British Columbia. The perception of conflict arose because Miller was on a permanent leave of absence from his timber industry (pulp mill) job, raising the question of whether Miller's judgment could be clouded in matters involving his past and likely future employer.

Indeed, it raised an even broader question: could this anticipated future personal benefit have had an influence upon the Minister's ability to balance the discharge of his more general duties as Forests Minister, especially since the crucial issues of the day involve fairness to non-timber interests (i.e., non-industry and non-government organizations, groups and individuals) and ensuring the application of true integrated management principles to forest management?

Miller's attitude and approach toward environmentalists and non-industry/non-government organizations has been highly adversarial. It may well go a long way toward explaining why BC's newly elected NDP government appeared so vulnerable on forest issues during its first year in power.

Within that first year, Premier Harcourt launched, under the direction of Stephen Owen, the Commission on Resources and Environment, a conflict resolution process explicitly designed to resolve forest use conflicts through a shared decision-making approach. At the same time, however, quite a different message was being sent by the Forests Minister. By inclination and experience as a union organizer in Prince Rupert, Miller is a combatant; the Harcourt and Owen approach, with emphasis on shared decision-making among all interests, is in direct conflict with those who would carry on the traditional adversarial-style battles. And make no mistake, the vision of a different approach to forest conflict resolution is a personal undertaking of the Premier. While there are a number of individuals within Cabinet, including those who continue to serve outside of Cabinet appointments as "back benchers," who enthusiastically support the shared decision-making approach, regrettably, majority support within the NDP is not there. It's a good idea in an ideal world, they muse, but impractical in the real world where ministers must get on with the business of representative governing, and can't abdicate their responsibly to govern by including opportunity for direct participation by citizens who will be affected by ministerial decisions.

In the real world of our British parliamentary system, the ministers prefer their splendid isolation from ordinary citizens. The democratic need of our pluralistic Canadian society is to empower people to achieve a much greater degree of participation in decision-making, rather than continuing to limit most interested people to "a kick at the can every four years." Prime Minister Mulroney got the message after his disastrous national referendum campaign. Sooner or later, provincial politicians who remain entrenched in old patterns must acknowledge that society has changed. The "paradigm shift" of public values related to the environment has radically modified the ability of politicians to carry on with methods that no longer achieve politically workable solutions.

Building a Better Participation Process

It is easy for a bureaucrat to say s/he was hired to look after the public interest and therefore need not consult effectively with the public. It is easy for a logging company executive to attempt to maximize the short-term interests of shareholders at the expense of everybody else. It is equally easy for an environmentalist to simply oppose logging as an answer to otherwise legitimate ecological concerns. The longer a dispute continues, the more extreme and entrenched are the positions of the parties. Forest disputes have become very entrenched.

Changing our way of relating to each other—from a competitive, adversarial system toward a shared decision-making approach—requires an act of will. Knowing we need to change is not enough. In order to succeed, we must have clearly in our minds the principles upon which to build the transition. If we are to succeed, we must identify the principles of administrative fairness upon which shared decision making is founded, and be prepared to challenge public participation situations that are patently unfair. There is nothing worse, nothing more discouraging or disempowering, than for government or industry to invite the public to participate, only for the involved public to discover later that high sounding principles are nothing but a cloak for a hidden agenda being pursued by one or more of the parties at the table.

The public has a right to demand that forest planning and decision-making processes achieve the goals they are designed to meet. Faced with a forest dispute, government creates public processes—

planning teams, boards, commissions and the like. More often than not, these processes take up inordinate amounts of time and resources, accomplishing little. From the moment when government or industry first announces the creation of a "consultation" process, the public should be capable of evaluating the fairness of the proposed process. Then, every step of the way, all participants in the process, participant constituencies, and the public generally, should be in a position to evaluate whether the process remains fair and equitable toward all legitimate forest interests.

Basic Principles for a Fair Process

The technical term used by the experts is "negotiated, interest-based, consensus building." The unique term that combines the intent of all three labels is called shared decision-making—an invention, to a significant degree, which was made jointly by the former ombudsman of British Columbia (1987-1991), Stephen Owen, and the former leader of Her Majesty's Loyal Opposition, Michael Harcourt. When the New Democratic Party came to power in late 1991, Harcourt and Owen put their idea into effect—and the permanent Commission on Resources and Environment was born.

The chief technical or process architect who then developed the concept into workable form was Craig Darling, a Brentwood, BC, lawyer with a background in community economic development. Use of the shared decision-making method (an approach, really, with many possible variations) challenges us to fundamentally rethink the way we organize ourselves to solve resources and environment problems. There is no one correct cookbook method of applying shared decision-making to a forest dispute. Considerable latitude is sometimes required to ensure the process meets the needs of the participants. However, the principles of fair process always remain constant—that is, in a modern democracy, citizens have a right to be meaningfully involved and to participate directly, fairly, and with a measure of equality with other more established sectors in the formulation and implementation of government decisions which affect their interests. With this in mind, the application of the principles of shared decision-making should be readily visible to process participants and their various constituencies. Achievement of these principles needs to be measurable in any forest "involvement," "participation," or "consultation" process that claims to be administratively fair to all who seek to participate effectively.

The 15 principles listed below are distilled from the experience of innovators in the shared decision-making approach—from people such as Gerald Cormick and Craig Darling—as well as from the experience of the people of BC as they attempt to resolve longstanding forest use disputes.

Principle #1: All sectoral interests shall be given a fair chance to participate directly in the dispute resolution effort. The process must be inclusive.

In forest disputes, this means that all unique perspectives, values, and beliefs are legitimate. The purpose of the process is to accommodate a diversity of interests—not to compromise some in favour of others.

Few, if any, of the multitude of Forest Service sponsored public consultation initiatives meet this basic test of administrative fairness. In the 1988 annual report of the Ombudsman of British Columbia, in a chapter entitled "Integrated Resource Management: The Issue is Fairness," then Ombudsman Stephen Owen wrote,

> Meaningful participation means that individuals, groups, local governments, or any other party with significant and legitimate interests will be recognized in the planning, implementation and conflict-resolution processes. It implies that their representations will receive careful consideration and will be accorded due regard consistent with the importance of the individual's interest. In other words, a duty is placed upon the decision maker, insofar as is reasonably possible, to appreciate fully the significance of and the foundation for the various individual or group interests.

Most public officials will argue that public involvement or consultation efforts don't involve shared decision-making, but are merely designed to solicit "input" from the public. In the purest technical sense, based on traditional criteria, they would be correct. The Ministry of Forests has, for many decades, signed contracts (licence agreements) with corporations such as MacMillan Bloedel. None of these contracts include an obligation for the licensee to plan for the use of its public tenure areas based on a shared decision-making approach. In fact, these licences don't even require the consideration of integrated resource management or the protection of non-timber values. These values, if considered at all, have been at the total discretion of the licensee. Public values and "input" are isolated outside of the legitimate (licence agreement) process.

As the public and the myriad number of organized groups concerned with non-timber values become aware of the provincial government's systematic legalized exclusion of their interests, the level of forest conflict will increase at an exponential rate. The Forest Service does not invite public sector interests to share directly in the responsibility of decision-making, precisely because ministry officials lack legal authority to take public concerns into account. The licence agreement is always paramount; input and consultation is always an add-on, without justification under law.

Shared decision-making, however, means just what it says. Sectoral interest participation is full and complete, covering all aspects of problem solving. This includes the opportunity for review of the very decision-making process under which the sectors were called together. As the Atlantic salmon issue involving the provincial Ministry of Agriculture and Fisheries illustrates, sectoral participants must have a right to examine and propose changes to existing bureaucratic procedures when they are perceived to be an obstacle to reaching agreement. In the Atlantic salmon example, the fisheries bureaucrats did not perceive a need to work with the Clayoquot Sound Sustainable Development process. The bureaucrats' existing decision process proceeded separately, but parallel to the Clayoquot Sound participation process. The "all values, all interest" Clayoquot process participants were not informed, not invited to participate in formulation of the Atlantic salmon decision, and told to mind their own business when word of the bureaucrats' secret deliberations eventually leaked.

To achieve legitimacy, a shared decision-making process must either supersede existing government processes related to matters under consideration, or explicitly seek the agreement of participants about what issues are to be excluded. If other dependent and ongoing government activities will not come under the scrutiny of the shared decision-making process (e.g. introduction of foreign fish species), then the parties to the process must be given the opportunity to consider if such prior exclusion will affect their ability to function.

In general, shared decision-making is a replacement for existing processes. Citizen groups require of government the modification of forest planning processes in ways which empower them to participate effectively. When the Ministry of Forests simply adds on a public input process, like wallpaper over a wall full of structural defects, while failing to amend licence contracts and legislation to legally empower citizens to share in decision-making, it engages in a grand deception.

Principle #2: The offer of sectoral participation shall be timely. All participants shall be invited to the negotiation early enough to participate in all substantive elements of the process.

Timing of the offer by government (or industry) to participate is an important indication of whether the offer is genuine.

The current Tree Farm Licence Management and Working Plan (MWP) public involvement process is a good example of where the timing fails the test of fairness. The Forest Service has even published a brochure which graphically discloses the undemocratic manner in which participation is denied until after the point where all relevant decisions are made and the outcome is irrevocable. Prior to the offer of public participation, the Ministry of Forests, in consultation with the licensee, agrees on management objectives, procedures, and research to be done. Prior to the offer of participation, the District Manager approves the MWP. Next the Regional Manager of the Forest Service "signs off" and approves the MWP. Then the Victoria-based Branch Directors of Timber Management, Integrated Resource Management, Recreation, and Range, "sign off" and approve the MWP. By this time some 21 months of a 22-month process is complete, and still the invitation for the public to participate has not been extended.

It is near the 21st month that a back and forth exchange of draft Management and Working Plans begins between the licensee and the Ministry of Forests headquarters staff in Victoria. Until the Ministry of Forests is satisfied that the MWP is to their liking, the draft is sent back to the licensee to make written corrections based upon written "deficiency letters" prepared by senior staff. Once this stage of draft

MWP review is complete, all Ministry of Forests staff are officially satisfied that the MWP is satisfactory and meets the government's requirements. Then, and only then, is the offer for public participation made known. During the last 30 days of a 22-month planning process, the public is invited to make comments on the plan. By now, of course, all significant planning steps are complete and finalized. The offer to participate is not timely.

Principle #3: Participants shall be given the opportunity to review and propose improvements to the decision-making process in which they participate.

Once the issue of who will be represented at the negotiating table is initially settled, it is time for the participants to decide among themselves how they will conduct their affairs.

This, too, in a shared decision-making process, is fundamentally different from traditional consultation processes. Rather than being told precisely how the process will evolve—where rules are drafted beforehand by the agency sponsoring the process, usually the Ministry of Forests or a forest company—the participants consider how best to set their own procedures.

Participants must have the right to assess whether the terms of reference are complete and appropriate to the issues which must be resolved. Deficient, manipulative processes are usually crafted in a way to prevent participants from dealing with vital elements of disputes. In the Clayoquot process, for example, government unfortunately decided it could require consensus of all participants at the table before allowing a concern to be raised and examined. If participants do not have a direct role in drafting the terms of reference and procedures, the table will usually be unable to deal with the more important issues.

Principle #4: Senior public officials shall demonstrate consistent support to ensure that civil servants participate and contribute fully.

If public servants are participants in the process, as is likely, then they must consider the process as a normal part of (not separate from) their routine decision-making responsibilities. Their role at the table must be as fully contributing participants. This means the government does not send in junior staff, or even untrained senior staff, if either lack the skills and the authority to contribute meaningfully to the table's efforts to envision a solution.

An equally important element here is that government agencies should work cooperatively so that the government can be represented at the table by only one participant. This is a new experience for government agencies and ministries, which generally compete for power, authority, and control of political agendas.

During the Clayoquot process, a specific negotiation among the parties resulted in an agreement between MacMillan Bloedel and the environmental sector interests to prepare a plan for a watershed called Tofino Creek. As part of this negotiated agreement, the Ministries of Environment and Forests agreed to act as co-chairs of the watershed planning process, and to use Community Watershed Planning guidelines (Appendix H) of the Ministry of Environment's policy manual as the basis for planning. These guidelines, from the Ministry of Environment, while primarily intended to address water concerns, apply a more holistic approach to forest planning. Simply put, the guidelines do not automatically assume that logging will take place, leaving the only unanswered question to be how to mitigate the impacts. Instead, these guidelines assume that a full range of watershed values is first inventoried. At about stage 12 in the guidelines, the opportunity arises to test a logging proposal against the documented forest values. Then, and only then, can a logging plan be considered, leaving open the possibility that if the plan does not protect all the identified values, it will not proceed.

In contrast, all the Ministry of Forests planning processes presume, first, the right of logging to occur. Then, other values are "fitted in," with mitigation of impacts to other forest values as one purpose of planning. Tofino Creek watershed is part of a Crown forest tenure and therefore, Ministry of Forests planning procedures apply. The Ministry of Environment has no legal standing to make decisions within such a forest tenure. But in the case of the Clayoquot process, normal Forest Service planning procedures were suspended as part of the terms of reference for the process. The concerned parties agreed during the course of the Tofino Creek negotiations to actually work within a planning framework which was different from normal

Forest Service practices. It was assumed, on good faith, that the government representatives, including those from the Ministry of Forests, would abide by the terms of reference negotiated for Tofino Creek, and that they had the full endorsement of their superiors within their respective ministries. It was assumed the bureaucrats would participate in the Clayoquot process in good faith, and as a consequence, treat the negotiated agreement concerning a new type of planning process for Tofino Creek with respect.

It was only several months later that participants in the Tofino Creek watershed planning process discovered that the Forest Service still intended to control the process in accordance with its long-established Local Resource Use Plan (LRUP) guidelines. In reality, the Forest Service participants to the negotiation (a) had no authority to abide by the Clayoquot process terms of reference, which clearly stated that existing Ministry of Forests planning processes were suspended, and (b) would not permit the more ecologically balanced Community Watershed Planning guidelines to be used in Tofino Creek as had been negotiated between the parties. Instead, with timber as the primary interest, LRUP guidelines were unilaterally imposed upon participants by the Ministry of Forests. As well, the co-equal chairmanship status between the Ministry of Forests and the Ministry of Environment quickly deteriorated back to control by the Forest Service.

The government participants in the Clayoquot process were not keeping their superiors informed and updated on negotiations, did not in reality have authority to negotiate agreements such as the one in Tofino Creek, and could not, as a consequence, implement the negotiated terms.

Principle #5: The representatives participating in negotiations shall ensure their respective constituencies are fully informed about negotiations on an ongoing basis.

All interests represented in negotiations must keep their constituents fully informed and involved in all aspects of negotiations. This is a responsibility which has seldom been fulfilled. For example, the Port Alberni Environmental Coalition participated, for a short time, in a local advisory committee intended to support and provide advice to the Port Alberni representative in the Clayoquot Sound negotiations. However, Port Alberni's representative, Mayor Trumper, ignored the advisory committee, did not attend its meetings and did not inform it of events occurring at the negotiations, according to Coalition spokesperson Judith Hutchison. As a result, the Environmental Coalition resigned from the Port Alberni advisory committee.

Principle #6: The definition of legitimate sectoral interests shall be sufficiently broad to include all interests (local, provincial, national and global) in forest land use.

Shared decision-making also assumes that the parties represented at the table are there to represent particular interests. And while it is natural that certain aspects of interest may overlap and constituencies develop, the separate parties are not intended to represent an aggregate of a single special interest. Yet this is precisely how the participant selection was developed for the Clayoquot process by the provincial government who convened the process. The lone individual interest was that of the District of Tofino.

This became self-evident over time as local government representatives from Port Alberni, Ucluelet, the regional district, and the IWA declared themselves satisfied with any agreement that met the interests of MacMillan Bloedel and Fletcher Challenge Canada, leaving Tofino sectoral interests (of which there were initially three) as the sole active participants in the Clayoquot process. All other participants openly declared that they had no interest to represent at the table, and that if they did have an interest, it would be represented by the two multinational logging corporations. Negotiations occurred only between Tofino and the two companies during the first year of the process. The remainder of the participants were passive. They had passed to MacMillan Bloedel and Fletcher Challenge Canada their proxies to do with as they liked.

In contrast, the community of Smithers recently created their own Community Resources Board. The community defined a Board member as a person who will predominantly represent one of the following resource values, thereby ensuring that all values are fully represented on the Community Board:

- timber production
- timber production for small operators
- preservation of natural ecosystems
- preservation of large tracts of wilderness, with limited access
- management of forest land resources to maintain habitat of hunted animal species and aesthetic quality of hunting environments
- trapping
- quality of fish habitat and the aesthetic quality of fishing environments
- subsistence, lifestyle and spiritual values
- tourism
- water quality for agriculture
- enhancement of recreation access and recreation facilities with minimum activity restrictions
- preservation of access for prospecting and mineral development
- dependent secondary industry
- intensive management
- preservation of aesthetic features of forest lands including landscapes and localized natural attributes
- preservation of historical and cultural features of forest lands

Principle #7: The provincial government shall ensure that the involvement process is fully recognized as the sole legitimate planning method.

All parties to a shared decision-making process must have sufficient incentive to participate wholeheartedly. In other words, the process must be recognized as the legitimate (normal) planning route from which decisions will eventually be derived, and government must make it very clear, through enabling legislation, that the shared decision-making process has behind it the full force of the law.

With the exception of some major areas of private forest land along old railroad land grant right-of-ways, almost all forest land in British Columbia is publicly owned. The vast majority of forest disputes concern public land. The provincial government, therefore, is in an extremely strong position to ensure the integrity of forest planning processes—if it so chooses.

The provincial government's commitment to the Clayoquot process was sufficiently weak to allow participants who came to the table possessing considerable traditional power to continue to exercise that power without regard to the principles of shared decision-making, which were agreed to as a condition of their participation in the process. Gerald Cormick, who held a workshop on the principles of shared decision-making for participants early in the first year of the mandate, said, in addressing an April 1992 conference in Victoria entitled "The Consensus Approach: Decision-making for the Nineties.":

> There have been some unfortunate experiences in BC where a variety of sectoral interests had invested tremendous amounts of time, come up with a consensus but those who are expected to implement it are off in Victoria and didn't even know what was going on. *The public officials hadn't invested in the process . . .*" [Author's italics.]

In the early 1980s, then Forest Service Chief Forester Bill Young authorized the creation of a planning team of interested parties to assess four options for the future of Meares Island, which also just happened to be in Clayoquot Sound. After two years of planning team consultation, MacMillan Bloedel was dissatisfied with the direction the options analysis was headed. The corporation withdrew from the planning team and unilaterally went to Victoria where it presented a status quo plan to the Social Credit Cabinet of the day. The need to uphold the authority of the province's Chief Forester was ignored by Cabinet. The need to uphold the integrity of the Ministry of Forests public consultation process was ignored by Cabinet. Instead, Cabinet approved an order to allow MacMillan Bloedel to proceed with logging immediately.

The international confrontation which followed is now history. Meares was not logged—and has not yet been, thanks to the intervention of the Nuu-Chah-Nulth Tribal Council and especially the Tla-o-qui-aht people. But the issue of one or more interests having the power to get its way by stepping outside the process remains unresolved to this day. Two quasi-shared decision-making reports (from the Strathcona Public Advisory Committee and the West Chilcotin Community Resources Board) have been on the desk of the Assistant Deputy Minister of Operations for the Ministry of Forests for over a year. He, too, has ignored the need to uphold the integrity of the planning process. As long as bureaucrats and logging companies can, with impunity, brush aside the work of shared decision-making processes, the public's confidence in forest governance will continue to erode.

Principle #8: Provincial legislation shall be in place to ensure that sectoral interests will participate in public consultation processes in good faith. The law shall ensure that, barring appropriate right of appeal, each participant views the consensus process as the "best alternative" to reaching an agreement to meet his or her needs.

The next best alternative to participants in a shared decision-making process cannot be personal access to the Minister's office with a private key. Historically, the traditionally enfranchised sectors—big labour, big business, big government—have had privileged access to the Ministry. What must be ensured for *all* sectors now is that shared decision-making is the best available option for accommodating each sector's interests. In the absence of this, some sectors will sit at the table and pretend to participate, only to cut a private deal later.

The public official is a participant—nothing more, nothing less. In the Clayoquot consensus process, the public officials gave their proxy support to the logging corporations in the same manner as local government and the IWA as mentioned above. First, public officials clearly understood that the shared decision-making process was not a normal part of routine decision-making for them, despite what the terms of reference for the Clayoquot process said. Second, and consequently, the public officials understood the need to maintain a posture which acknowledged that real planning and decision-making was simultaneously occurring elsewhere. This meant the public official could not endorse, support, explore or envision problem-solving alternatives that in any way appeared to conflict with the status quo. The public official was not, therefore, empowered to participate.

Principle #9: All relevant information shall be shared equally among all participants, in a timely fashion. Provincial law shall ensure that all participants will share information vital to the process.

The old adage is true—information is power. In the Clayoquot process, timely information necessary to assist participants to envision options for creating sustainable forest use was withheld. Early in the process, some participants asked that a satellite or aerial photo mosaic be provided so that the extent of remnant old-growth areas could be seen. Yet in spite of the presence at the table of three deputy ministers, the District Manager of the Ministry of Forests and other government officials (including five paid facilitators), the data was not provided for more than six months. What none of the bureaucrats were willing to tell the participants was that the requested photo mosaic was already compiled and sitting in the board room of the District Office of the Ministry of Forests in Port Alberni. Had the local District foresters been empowered to facilitate the resolution of participant information needs, the issue could have been resolved in a matter of hours, not months.

Much later, after nearly three years of experience with the "shared decision-making" method, the NDP's new Minister of Forests, Dan Miller, transferred control of one of two major forest licences in the study area without consulting participants in the process. In addition, fisheries officials continued to plan long-term and potentially hazardous modifications to a major freshwater lake through introduction of Atlantic salmon—again, without consultation with participants in the process. MacMillan Bloedel unilaterally wiped out hundreds of community jobs, without consulating participants in the Clayoquot "shared decision-making" process.

Principle #10: Provincial law shall ensure a relative balance of power among participants. Provincial law shall ensure honest and committed participation in the consensus process.

Is participation in the process better than the next best alternative for each participant? Use of the shared decision-making approach presumes a relative (not absolute) balance of power among sectoral interests. Each must have an incentive to participate willingly. If MacMillan Bloedel, as it did with Meares Island, can simply walk away from the table and get what it wants, no incentive to negotiate exists. If public officials can continue to allocate natural resources without consulting the shared decision-making participants, there is no incentive for the bureaucrats to participate as equals with those representing the non-contractual public interests (fish, wilderness, recreation, community forestry, tourism, etc.), who also have a legitimate right to an equitable place at the negotiating table.

The very nature of our system of governance tends to preclude the successful use of shared decision-making when a public (non-contractual) interest is involved. In contrast, a contractual interest (a timber licensee such as MacMillan Bloedel) enjoys extraordinary rights under law which can be exercised independently of other participant interests. In short, logging companies with contractual rights to timber licences are at a distinct advantage over all others who seek to represent public interests in fish, wildlife, wilderness, water quality and quantity, recreation, fisheries, cultural and spiritual values, heritage, and so forth.

In Canada, British parliamentary tradition allows that the Crown represents the public interest when disputes arise over non-contractual natural resource interests. In contrast, citizen groups and individuals in the United States have the right to use the courts to protect non-contractual natural resource interests.

It is this right to court access that has tended to balance power among the sectoral interests in forest use disputes within the United States. There, having reached a virtual stalemate in court battle after court battle, the parties may decide to participate in a shared decision-making process as relative equals because their next best alternative course of action (the courts, lobbying, etc.) no longer appears attractive, and because they have achieved a balance of power (of sorts) between the disputants due to the levelling effect that access to the courts tends to give the less powerful.

If we choose to open up access to the courts for citizen interests in natural resource disputes, a balancing of relative power may occur, similar to that found in the United States.

However, perhaps our best alternative opportunity is to envision a different course through enabling legislation establishing principled negotiation and consensus processes as a "legal right" of sectoral interests. Having enacted such legislation, the possibility exists to build an approach to joint problem solving which could lead us toward a model of stewardship for our forests unparalleled elsewhere in the world.

Principle #11: Participants shall be able to envision options for other participants to consider. The context for problem solving shall be sufficiently broad to allow participants to bring all relevant social, and/or economic, and/or environmental factors to the table for consideration.

Tradition holds that public consultation processes be defined with quite narrow terms of reference. Many important social, or economic, or environmental factors are, by this definition of public consultation, excluded from consideration by participants. Gerald Cormick, at the Victoria conference in April 1992, said, "I've never been involved in a dispute yet that I didn't have to make bigger in order to settle. This is in contrast to the normal process of conflict resolution in which it is the convenor's objective [usually the Ministry of Forests or a timber licensee] to narrow the issues . . ." Cormick is the recognized "father" of the successful application of the process of consensus to large-scale public environment/resource disputes.

Principle #12: The process shall clearly define how the recommendations, findings or decisions of the participants will be implemented by decision-makers or public officials.

From day one of a legitimate shared decision-making process, participants must develop a common understanding of how the results of their negotiations will be used. Public officials must reach agreement with the participants, from the very beginning of the process, on the mechanism by which the results of the negotiations will be implemented, approved or reviewed.

In our earlier examples which cited the recent Strathcona and Chilcotin processes, the public decision maker at the executive level of the Ministry of Forests (to whom the findings of the processes were presented) simply undertook no commitment to do anything. He accepted the reports and henceforth ignored them.

Principle #13: The process shall provide for a resumption of negotiations to deal with new disputes that may arise later.

The Clayoquot process was managed by government officials in a manner that led participants to conclude erroneously that their final report would, henceforth, "establish conditions for all agencies working in Clayoquot Sound." By "all agencies," the participants inferred that all individuals, organizations and government agencies would be required to abide by their conclusions. Clearly the participants lacked knowledge of the very fundamentals upon which their process was supposedly based. Experienced dispute resolution practitioners acknowledge that they have never permanently settled a single conflict. Cormick puts it this way: "I've never resolved a conflict in my career of dispute resolution. Never! I settled lots of disputes, but the basic conflicts don't go away. The basic conflicts are there and will be fought again."

The conflict resolution process must build into any agreement precisely how the parties intend to handle future disputes. Presumably, having once been successful in settling their disputes, the task of settling future disputes will not be so difficult. Building into agreements the means to address future disputes is crucial.

Principle #14: The process shall provide for experienced facilitators, as required. The process shall provide for the participants themselves to choose their facilitators. The facilitator shall assist participants to explore fully all feasible ways and means of meeting their interests.

Facilitators are helpers, but the participants must design their own process within a framework of shared decision-making principles. Mediators are also helpers, but again, it is the participants who must design their own process within the framework of sound participatory principles. Depending on the size of the process and its needs (a reflection of the complexity of the dispute), there may be several helpers, or there may not be any. The role of mediator may at times be intertwined with that of facilitator. Each role is characterized by these phrases: a listener, a trust builder, a bridge builder, an initiator of communication, and an information provider—all apply.

In the Clayoquot process, government initially employed five facilitators, including one forester. None appeared to do more than arrange meeting rooms and ensure the presence of coffee during scheduled breaks. This is not the primary purpose of facilitators, although they may indeed perform these functions in the course of carrying out their primary duties.

Particularly in environmental disputes, facilitators more often than not must fulfill an active interventionist role. However, the meaning of the term "interventionist" must be carefully defined. An interventionist role is required when the parties to the dispute lack technical information, are inexperienced in negotiation, or are unaware of possibilities that may aid in the discovery of substantive agreement among the participants. An aggressive process participant, such as a major licensee or a union representative, may instinctively object to an interventionist role on the part of either a facilitator or mediator. Both the timber licensee and the union representative are likely to be highly skilled in aggressive bargaining techniques due to their decades of experience in facing each other across the table in more traditional labour-management disputes. But other participants at the shared decision-making table, especially those representing non-contractual public interests, are likely to be inexperienced. The mediator and facilitators must, if fairness is to be achieved, assist the inexperienced participants to develop necessary skills and to obtain access to information. Both the facilitators and the mediator have a crucial responsibility to "level the playing field."

The field of forest research and knowledge has expanded at such a rapid pace in the last few years that few potential participants will enter into negotiations with all of the information they need to envision many of the feasible elements of a dispute settlement package. A facilitator must not only be a good listener, but must also be sufficiently knowledgeable to put participants in touch with information that can assist them in exploring fully all feasible ways and means of accommodating their interests.

An alternative to having line ministry or agency bureaucrats at the table would be to use their considerable technical talents as support personnel, or as expert information facilitators in a "second-tier" role. Once representatives of all community interests are placed directly at the shared decision-making table, the non-elected public officials could serve in a support capacity. This option presumes a much greater working knowledge of shared decision-making methods on the part of the bureaucrats than currently exists. It is important to remember that although non-elected public officials, in whatever capacity—at the table, or in a "second-tier" support role—have special responsibilities related to administering ministry acts, regulations and policy, they have no special status.

Since most traditional planning processes are convened by either the Ministry of Forests or a logging corporation, their role as both participant and convenor is confused, and detrimental to the achievement of successful agreements. When one of the principle participants imposes the facilitators upon the process, the facilitators tend implicitly to pursue the agenda of their employer. Thus, a facilitator and/or mediator who is appointed by one of the participants can be a good indicator of a sham consensus process.

Principle #15: If a mediator is to be involved, the participants shall have the right to choose the mediator. If the dispute resolution process is chaired by one of the participants, the decision on who will be the chairperson shall be made by mutual agreement of all the participants.

The old concept of planning meant experts talking to experts on behalf of everyone else who didn't know any better. The whole nature of planning has changed. Planning is now about recognizing the legitimacy of diverse values, and accommodating that diversity. When all of the participants realize this, it changes the nature of participants' commitments from mistrust to cooperation.

A mediator trained in shared decision-making methods is an individual capable of working with a diverse array of participants to help them achieve substantive agreement which satisfies their interests. The mediator in environmental or forests disputes is a specialist in a new field. Practitioners of labour mediation, for example, may not necessarily make a successful transition to environmental mediation. Shared decision-making is a new and unfamiliar concept. Shared decision-making means that non-contractual public interests are given a legitimate place at the negotiating table.

Working Towards Public Participation

The Ministry of Forests already engages in a myriad of public consultation processes. Yet none approaches the "standards of fair process" set forth in this chapter. Currently, the role of non-governmental organizations, interest groups, individuals, and others without legal contractual relationships with the Ministry, is limited to advisory status only.

In the United States during the past decade, principled negotiation (consensus) has been demonstrated to be effective, efficient and appropriate for resolving environmental conflicts. The unanswered questions for the Ministry of Forests in British Columbia are these: Why has the Forest Service made such very limited use of this method of dispute resolution? Why does the Forest Service not officially recognize this method of dispute resolution in Ministry policy? Why is there an absence of legislation, regulations and policy supporting this approach? Fear of losing control appears to be the single largest impediment to the use of shared decision-making in the Ministry of Forests.

The Ministry has, for decades, wielded an immense discretionary power over the use of our public forests. So extreme is this discretionary power that the RCMP, in an investigation (several years ago) related to unpaid timber cutting fees, found that their charges could not proceed because the Forest Service interpreted its stumpage policies to yield exactly opposite conclusions from day to day and from licensee to licensee. Grown accustomed to this extraordinary power, the Ministry of Forests may be unwilling to engage in any process it perceives as mitigating its legal autonomy. Various provincial commissions and inquiries, including the 1986 Wilderness Advisory Committee, recognized this problem. Establishing shared decision-making as a "legal right" may be the only way to modify the behaviour of the Ministry of Forests.

Considerable work has been done by the British Columbia Roundtable on the Environment and Economy in the area of shared decision-making. In May of 1992, the Roundtable announced its support for a vision of shared decision-making which included the creation of Community Roundtables. Each community which chooses to adopt such an approach, as several have already on a voluntary basis, would likely follow the recommendations of the Roundtable which advocates the use of a principled approach.

If communities begin to evolve some experience in this method of governance, the need for the Ministry of Forests as an all-encompassing authority over forest use diminishes. What politician, when faced with consensus from constituents, will say no?

The means to rehabilitate and redefine "public input" and "public consultation" can be found to a very large degree in application of shared decision-making principles in forest dispute resolution. The application of this approach to forest disputes means democratically fair public participation processes. Fair process enhances the likelihood of participants achieving fair agreement. Fair agreements lead to greater social and economic stability.

Let us be clear on the conceptual difference between planning and dispute resolution—*there is no difference.* Forest planning processes are, by another name, also forms of dispute resolution. The stages of a credible forest planning process bear striking similarity to those of successful dispute resolution methods. The prime difference is that forest planning is traditionally done by bureaucrats sitting down with forest licensees such as MacMillan Bloedel and Fletcher Challenge. The concept of shared decision-making is nothing more than, first, adding a set of administratively fair procedures to guide the evolution of forest planning, and second, adding to the process all participant interests rather than leaving the decision-making or planning effort exclusively in the hands of public officials and licensees.

We have outgrown many of our institutions. Our representative parliamentary democratic tradition has not kept pace with changing public values. On any given day in British Columbia it might be possible to tally up to 300 separate "public input" or "public consultation" processes of one form or another that are sponsored by the Forest Service. Each process invites public input. Each process claims it will take public concerns into account. Yet, of all these processes, not one of them can be said to include the public as a direct participant in either forest planning or forest decision-making.

Gordon Sloan, mediator for the Vancouver Island negotiation for the Commission on Resources and Environment acknowledges that never has shared decision-making actually been applied to public environmental dispute resolution in British Columbia. With the creation of the Commission headed by former Ombudsman Stephen Owen, British Columbians will be afforded their opportunity to invent a better way of reaching agreement. Opposition to changing our ways will be encountered, but the nature of the world today makes substantive change in our approach to decision making an imperative. The time has come to apply shared decision-making principles in forest dispute resolution.

Postscript

In November 1992, the Commission on Resources and Environment began the process of regional negotiations on Vancouver Island. The objective was to provide Commissioner Stephen Owen with consensus recommendations on a land use strategy which would accommodate the interests of all groups or sectors involved in the forest debate in British Columbia. The Commission modelled the negotiations upon the concept of shared decision-making. The sectors include Recreation, Direct Forest Employment, Tourism, Local Government, General Employment, First Nations, Conservation, Forest Sector (Majors), Youth, Provincial Government, Forest Sector (Independents), Social and Economic Sustainability (Share and local economic interests), Mining, Agriculture, and Fisheries.

Few expected the task to be easy. However, it quickly became apparent that the problems would not arise due to an inability of widely diverse community interests to work together cooperatively. Quite the contrary, in fact. The only sectors with any difficulty at all were those with a perceived vested interest in the traditional, back door approach to reaching agreement with the provincial government on land use issues.

The first accomplishment was creation of a mutually agreed-upon process and procedures document to guide the work of the Table. It reads, in part, "Participants at the table acknowledge that their negotiation is a shared decision-making process. Shared decision-making means that on a certain set of issues, for a defined period of time, those with authority to make a decision and those who will be affected by that decision are empowered to jointly seek an outcome that accommodates rather than compromises the interests of all concerned." Next came the drafting of a Vision Statement to the year 2020 for

Vancouver Island. All sectors agreed, or thought they had, until the Forest Sector (Majors) were discovered to have been lying low, taking a wait-and-see attitude that resulted in the first real challenge to the Commission's shared decision-making process.

Having apparent agreement of all sectors on a Vision Statement, with only a bit of fine tuning remaining, the Forest Sector (Majors) came to the March 25, 1993, meeting of the regional negotiation Table, in Campbell River, prepared to test the commitment of the sectors—to both shared decision-making and their resolve to craft a social, economic and environmental transition strategy to move us from the status quo of present land use practices toward "sustainability." History may well identify this event as a significant turning point in the "attitude" of Forest Sector (Majors) toward the portion of British Columbia Society which has been clamoring for years (without success) to become partners in forest land use decision-making. Up to this point, with the Ministry of Forests controlling fully 86 percent of the land base of British Columbia, and the Forest Sector (Majors) the only legally entitled "partner," it is perhaps understandable that the Forest Sector (Majors) developed an attitude of exclusive rights. The following side-by-side comparison of the two Vision Statements well demonstrates the source of the problem and the willingness of society in general (as represented by the sectors interests at the Table) to envision a far more inclusive approach to accomodating each other's hopes and interests.

The Forest Sector (Majors) demonstrated their presumption of an exclusive right to control process and decision-making with their response to the Table's Vision Statement. Had the Majors been accustomed to accommodating the interests of the other forest users, their approach would have been quite different. Nevertheless, the experience was positive, although somewhat traumatic for many sectors. Having tested the resolve of the Table, the Forest Sector (Majors) learned what Prime Minister Brian Mulroney already knew about the need to accommodate the interests of others in a pluralistic Canadian society. The commitment of the Table sectors—and society in general —to a shared decision-making approach that accomodates the interests of all, is very real. The Majors quickly retreated from their "position" in the face of unity from all the other sectors, and agreed to craft a final Vision Statement which accommodated the interests of all sectors.

Two Versions of a "2020 Vision for the Vancouver Island Region"

Table Draft from March 11–12, 1993	Forest Sector (Majors) replacement vision dated March 22, 1993.
(**Bold** indicate portions proposed to be struck out by Forest Sector (Majors))	<u>Underlined</u> portions are Forest Sector (Majors) additions

We place the highest values on the well-being of the land, communities and people of our region and seek to sustain these values for all future generations. We seek balance in all things, between—wild and developed lands; environment and economy; **population and resources;** rural and urban communities; standard of living and quality of life; **self-sufficiency and interdependence;** ecological and social justice; common good and private interests.

We place the highest values on the well-being of the land, communities and people of our region and seek to sustain these values for all future generations. We seek balance in all things, between—wild and developed lands; environment and economy; rural and urban communities; <u>social standards and market competitiveness;</u> standard of living and quality of life; ecological and social interests; and common good and private interests.

We recognize the responsibilities **we bear** for the protection of our **global** heritage through the stewardship **and husbandry** of the environment. We seek the conservation and **restoration** of all ecosystems and their biological diversity—forest, aquatic, mountain, grasslands, croplands and wetlands, **which form the basis of a sustainable economy.**

We recognize our responsibilities for the protection of the natural heritage through the stewardship of our environment and for the provision of <u>goods and services globally shared. We seek conservation of ecosystems and their biological diversity—forest, aquatic, mountain, grassland, croplands and wetlands.</u>

We recognize the rights to habitat of all communities—human, plant and animal—which are united within the web of life. We assert the right of humans to earn a living from the bounty of the land based upon sustainable practices **that do not exceed the capacity** of natural ecosystems to restore themselves. We seek meaningful and diversified work which brings pride to workers of all kinds. The right to work is based upon work done right, which means full utilization of the natural resources extracted, yielding maximum employment opportunities.

We assert the right of humans to earn a living from the bounty of the land based upon sustainable practices. We seek meaningful and diversified work which brings pride to workers of all kinds.

Two Versions (continued)

Table Draft from March 11–12, 1993	Forest Sector (Majors) replacement vision dated March 22, 1993.
(Bold indicate portions proposed to be struck out by Forest Sector (Majors))	Underlined portions are Forest Sector (Majors) additions

We are committed to social justice and equity for all the Island's people, particularly youth and the aboriginal people, in the distribution of all socially valued resources.

We support participatory democracy guided by shared decision making.

We support participatory democracy guided by shared decision making **in which there is an equitable distribution of power. We also share in the impact of a transition to sustainability.** Workers and local communities are full participants in planning and decision making affecting them.

Workers, local communities, and industries are full participants in planning and decision making affecting them.

In all of this we maintain a strong sense of place and continuity, through time, expressed in a love of our beautiful island home which provides us with our spiritual, recreational, and physical needs.

In all of this we maintain a strong sense of place and continuity, through time, expressed in a love of our beautiful island home which provides us with our spiritual, recreational, and physical needs. We believe in a guarantee of sustainable forests and the shared, careful use of all resources for the socio-economic well being of present and future generations. We are conscious of the social and economic obligations to the people of Vancouver Island and the reliance that is placed on the healthy environment and economy. We recognize that our economy is highly dependent on the world economy. Our ability to compete in world markets will define the economic strength of our society and our ability to deliver social programs. We are committed to ensuring that social, recreation, cultural, ecological and economic values are integrated into the planning processes on Vancouver Island. To this end, we can employ principles of sustainability and maintain the economic contributions to the communities, province, and the Nation.

And what of the politicians? Shared decision-making is perceived by some as a threat. To the extent that shared decision-making openly and fully informs all sectoral interests of feasible options and alternatives, then the politician should embrace the opportunity to build wider consensus among traditionally competing interests. Yet some still argue that shared decision-making threatens our parliamentary democracy. Those who make this case are concerned only about the possibility of the constraint which the shared decision-making approach places upon their ability to make discretionary decisions without the advice and consent of the governed—or to cut backroom deals with their favourites.

On Earth Day, April 22, 1993, in Port Alberni, Commissioner Stephen Owen issued a Public Report and Recommendations on "Issues Arising from the Government's Clayoquot Sound Land Use Decision." Earlier in the month, the provincial Cabinet announced the boundaries for the forest to be protected, those to be under special management, and those for integrated resource use in Clayoquot Sound. The decision failed measurably to accommodate all who have legitimate interests in forest lands, including the Tla-o-qui-aht, Ilesquiat, Ahousahat, Toquaht, and the Ucluelet First Nations. Owen's report did not challenge the right of Cabinet to set the Clayoquot Sound boundaries for working forest, special management and protected areas. But what the Commissioner did challenge, on behalf of the sectors participating in the regional negotiations, was the failure of Premier Harcourt to tell us how forest practices were going to be improved as promised in the Premier's announcement of the Clayoquot decision. The Cabinet decision was widely perceived to express a failing commitment and a diminished resolve on the part of the Harcourt government to support the shared decision-making negotiations underway on Vancouver Island.

Owen's report notes that Premier Harcourt has not yet honoured promises on 1) a jobs and economic transition strategy for forest workers made necessary by structural change in the industry, 2) a Forest Practices Act, 3) biodiversity guidelines, 4) interim management guidelines, 5) criteria for a protected areas strategy, 6) a framework for First Nations participation in CORE, and 7) a compensation policy. That's a lot of unfulfilled promises, all of which Premier Harcourt said would be in place to support Commissioner Stephen

Owen's work to end the valley-by-valley forest conflict in BC. With all of these unanswered questions, Commissioner Owen has been forced during this past year to do most of his Commission on Resources and Environment work in the dark. Little wonder that Owen felt compelled to say what he said, and do what he did, on Earth Day. Yet, what a departure his forceful report is from the approaches of the past. Having a Commissioner mandated to seek accommodation for all who have an interest in land use decision-making really is important. That Stephen Owen was formerly a provincial Ombudsman only improves the chances of honesty and integrity in what his report says.

The Commissioner's legislative mandate, written and enacted by the current NDP government, gives Commissioner Owen the authority to report and recommend to the provincial Cabinet on resource and environmental issues, and on the need for legislation, policy, or land use practices. The Act gives Owen statutory independence from government, and his wide legislative powers include the coordination of government land use initiatives to ensure that government actions are both effective and balanced. Above all else, Owen's credibility rests upon a duty of impartiality.

The government's Clayoquot Sound decision promised to set and ensure new, world-leading timber harvesting practices and standards, an important issue being negotiated at the Vancouver Island regional table. But Owen correctly notes that Premier Harcourt didn't provide any details of those promised practices and standards. Owen recommends that prior to issuing new or revised cutting permits, Harcourt should provide specific details of the new practices and standards, explain how quickly the forest industry must comply, and how the public will participate in the development and approval of revised management plans and the issuing of permits. As well, Owen wants Premier Harcourt to spell out the penalties that will be imposed for non-compliance.

Owen is saying that Premier Harcourt's Clayoquot decision must measure up to national and international environmental standards. The report asks "government to designate Clayoquot Sound as a UNESCO Biosphere Reserve," and to ensure the as yet undefined new forest practices and standards are "directly evaluated against the Canadian obligations under the UNCED Biodiversity Treaty signed at Rio de Janeiro in 1992." Also included is support for having a federal

Model Forest Program. "These approaches would all help to ensure international recognition of the sustainability of resource use and environmental protection in the region and the benefit of the best possible science and technology," Owen said.

Some Cabinet ministers never believed in CORE's more democratic approach to land use decision-making, and would be quite happy to see an end to Stephen Owen's mandate. Shortly after the Cabinet announcement of the Clayoquot decision, Forests Minister Dan Miller publicly speculated that perhaps the goals of the CORE process were too lofty for us. Miller, it seemed, was suggesting greater comfort with the traditional approach of unfettered discretionary power to cut backroom deals with political favourites.

Premier Harcourt, fortunately, has to weigh the broader public interest. There is a real appetite among an overwhelming majority of British Columbians to make the CORE consensus building process work and thereby end the era of polarized land use conflict. Harcourt's choice is simple. Support the few anti-democratic voices in both his Cabinet and the forest industry who cling to the political and forest management status quo of a bygone era, or support CORE's shared decision-making process and reaffirm Stephen Owen's mandate to bring us together to craft joint solutions that fairly balance the interests of all.

Acknowledgements

Most communities in British Columbia are surrounded by forests. Since 1977, I have had the privilege of working with many people living in quite a number of these communities—trappers, ranchers, carpenters, doctors, pulp mill and woods workers, logging contractors, fishermen, entrepreneurs, and many more. Our bond and mutual interest is the forests, and a heartfelt desire to bring about a better approach to forest stewardship. But there are two communities that have taught me the most. They are Ucluelet and Tofino, on the west coast of Vancouver Island. There, I have learned to understand the real meaning of community and commitment. The yarn that I often tell to people is that on any day of the week, any week of the year—long before the Meares Island Planning Team was established back in 1984—there has been at least one meeting each and every day dealing with some aspect of deciding the future of Clayoquot Sound.

These two communities have continued to struggle toward development of a consensus and agreement on a practical definition of forest stewardship. I have watched as Tofino learned how to strengthen ties with the Ahousaht, Hesquiaht, and Tla-o-qui-aht people whose territories encompass Clayoquot Sound. I wish to acknowledge all the people of these communities whose dedication to improving forest stewardship had been an inspiration.

To Maggie, I acknowledge unceasing support, without which this work—and much more—would not have been achievable.

Further Reading:

Axerod, Robert. 1984. *The Evolution of Cooperation.* New York: Basic Books.

Bidol, P. and others. 1986. *Alternative Environmental Conflict Management Approaches: A Citizen's Manual.* Ann Arbor: Environmental Conflict Project, University of Michigan.

Bingham, Gail and Leah V. Haygood. 1986. "Environmental Dispute Resolution: The First Ten Years." *The Arbitration Journal,* Vol. 41, pp. 3-14.

Bingham, Gail. 1985. *Resolving Environmental Disputes: A Decade of Experience.* Washington, DC, Conservation Foundation.

Canadian Bar Association Special Committee. 1988. *Aboriginal rights in Canada: An Agenda for Action.* Ottawa: CBA.

Commission on Resources and Environment. 1992. *Report on a Land Use Strategy for British Columbia.*

Darling, Craig. 1991. *The Clayoquot Sound Process—An Analysis.* University of Victoria Dispute Resolution Centre.

Deutsch, Morton. 1973. *The Resolution of Conflict: Constructive and Destructive Processes.* New Haven: Yale University Press.

Fisher, Roger D. and William Ury. 1981. *Getting to YES: Negotiating Agreement Without Giving In.* Boston: Houghton Mifflin.

Fisher, Roger D. & S. Brown. 1988. *Getting Together: Building a Relationship That Gets to Yes.* Boston: Houghton Mifflin.

Goldberg, Stephen B., Eric D. Green and Frank E. A. Sander. 1985. *Dispute Resolution.* Boston: Little, Brown and Co.

Hoffman, Ben. 1990. *Conflict, Power, Persuasion: Negotiating Effectively.* North York, Ont.: Captus Press Inc., York University.

"How to Evaluate Fairness in Forest Dispute Settlement Processes (A Digest of Writings Dealing with Environmental Dispute Resolution, Mediation, and General Fairness Principles)." *Forest Planning Canada.* Vol. 8 No. 3, May/June 1992.

Laue, James H. (ed.). 1988. "Using Mediation to Shape Public Policy." *Mediation Quarterly*, No. 20. San Francisco: Jossey-Bass Inc.

Mansbridge, Jan. 1983. *Beyond Adversary Democracy*. Chicago: University of Chicago Press.

Raiffa, Howard. 1982. *The Art and Science of Negotiation*. Cambridge: Harvard University Press.

Schelling, Thomas. 1960. *The Strategy of Conflict*. Cambridge: Harvard University Press.

Susskind, Lawrence and Michael Elliott. 1983. *Paternalism, Conflict, and Co-Production: Learning from Citizen Action and Citizen Participation in Europe*. New York: Plenum.

Volpe, Maria R. and Thomas F. Christian. 1984. *Problem Solving Through Mediation*. Conference Proceedings, Dec. 1-2, 1983. New York: American Bar Association (Dispute Resolution Series no. 3).

A Global Context for British Columbia

Patricia Marchak

WOOD PRODUCTS HAVE BEEN IN THE GLOBAL MARKETPLACE throughout the history of trade relations, but until the late twentieth century they were produced only where nature provided the trees. Except for processors of decorative hardwoods, the forest industry was located, even two decades ago, in the northern coniferous regions; wood products were sold by northern companies mainly to other northerners.

As Japan developed industrial strength and the newly industrializing countries of Asia developed a demand for newsprint, new markets emerged. Then, in the 1970s and 1980s, new technologies gave life to new kinds of trees in southern climates. These are now becoming sources for industrial wood and paper, and the forest industry is no longer the preserve of the north.

The Creation of a Global Forest Industry

Decline of Traditional Softwoods

From about the middle of the nineteenth century to the end of the 1970s, the making of wood products from softwood forests was a central economic activity in northern Europe and North America. In the mid-1980s, nearly three quarters of all industrial wood (72 percent) was still grown in the northern temperate climates. The United States and Canada produced a third of this; Europe, about 20 percent; the USSR, about 18 percent. The remainder (28 percent) was produced in China (6 percent), other Asia–Pacific regions (6 percent), Brazil (4 percent), and elsewhere in smaller proportions.

North American forests, however, are in decline. They have been overcut and inadequately replanted. Since traditional softwoods take between 80 and 400 years to reach maturity, private investors are not eager to invest in reforestation. Instead, they have invested in the development of technologies to utilize tree species hitherto unsuitable for pulping and construction, and superior to softwoods in growth time and overall management costs.

Sweden has not suffered the same decline because it instituted reforestation and afforestation programs at the turn of the century. The forest cover there is now double the level of 1900. However, with expanding markets, discussed below, domestic fibre sources are no longer adequate, and Swedish companies are seeking new fibre sources. This is likewise true of Finland, which has had less reforestation and is short of domestic fibre sources. In addition, much of the European forest has been affected by acid rain.

Development of New Fibre Sources

The new producing regions include northern New Zealand; Tasmania in Australia; Indonesia, Malaysia, Thailand and some other parts of Southeast Asia; parts of Africa; Portugal; Spain; Brazil; and Chile; with some potential in Argentina and Venezuela; plus—a big plus—the southern United States. American companies based in Oregon and Washington were preoccupied for a while with the hooded owl that had captured the hearts of environmentalists. The far greater threat turned out to be more competitive fibre sources in Alabama where extensive pine and hardwood plantations are now coming on stream.

What these countries and regions produce are hardwoods, both tropical and temperate varieties, especially including eucalyptus; various new pine species; plant fibres, especially kenaf; sugar bagasse; and various other vegetable fibre sources.

Radiata Pine

Companies in New Zealand, Australia and the southern United States developed the radiata pine (known as Monterey in the United States) during the 1960s and 1970s. In New Zealand, native forests have been burned and bulldozed to make way for radiata pine plantations. These can be grown in relatively short cycles (about 30 years) and have high productivity per hectare. During the 1980s, one million hectares per year—much of it in pine—was planted in the southern United States.

Chile is the most promising source of new radiata pine fibres, with superior growing conditions and extensive afforestation projects already under way, and a reported 100,000 hectares planted by the late 1970s. Caribbean pine is becoming a major new fibre source in Central and South America. Ellioti pine is being grown in South Africa and Brazil. In Canada, and now in Sweden, lodgepole or jack pine is being harvested.

Tropical Hardwoods

Tropical hardwoods have become major sources of wood fibre for pulp, paper and paperboard only since the mid-1970s. In 1971, when the Paper Industry Corporation of the Philippines (PICOP) was founded in Mindanao, the utilization of local hardwoods was largely experimental. Since then, with genetic experimentation and extensive trials on all tropical woods in the Philippines, it has been established that local woods, either alone or in combination with softwoods brought in from elsewhere, are viable sources for high grade papers and paper boards.

Similar experiments have been conducted in other tropical regions—Brazil, Cameroons, Indonesia and elsewhere—and plantations have been established. Most of these trees can be grown in under ten years, and the yield per hectare is greater than for softwoods. Yields from intensively managed plantations in the tropics are estimated to be up to ten times greater than those in temperate zones. Further, in some regions—notably northern Brazil—the plantation acreage is enormous; in the Aracruz project it is reported to be 1.6 million hectares (an area half the size of Belgium).

These plantations are not tropical forests in the fashion designed by nature. Natural tropical forests are expensive to log because of their density and the extraordinarily rich mixture of woods, of which many are unsuitable for pulping. More economical from the perspective of large logging operations are plantation forests, where the jungle variety is eliminated and only those trees that can be transformed into high grade pulps are seeded. The view of the vice-president of Indah Kiat, a major pulp mill in southern Sumatra, provides a succinct expression of this:

We have 65,000 ha now; we are now in the process of getting conces-
sions for another 65,000. Basically we are looking for forest which can
be clear cut and replaced with eucalyptus and acacia.

Eucalyptus

Australian eucalyptus trees have been planted on other continents
since the late eighteenth century, usually for decorative purposes.
Plantations were seeded in the Mediterranean basin at the end of the
last century, intended to produce construction wood. Pulping technology changed from sulfite to sulfate methods over the 1960s, and then
the range of options increased to include several chemical and thermal
mechanical methods in the 1970s and 1980s.

The new techniques allowed companies to utilize a wider range of
fibres. Eucalyptus was found to be an ideal source. The advantages
included resistance to insects and fungi, and growth cycles of between
ten and twenty years. It produces over twice as much pulp wood as
pine, and is easier to cut and transport. Several harvests can be obtained from each stump.

Companies introduced plantations in warm climates well served
by fresh water. Seven pulp mills have been constructed in Portugal
since 1960, using eucalyptus as their fibre source. Eucalyptus plantations have also been established in Chile, Morocco, Spain and other
warm countries. The *eucalyptus grandis* can be harvested in seven years
for pulpwood, and the stumps then sprout new stems to be harvested
again on seven-year cycles. *Eucalyptus globlus* is now touted by industry
experts as the best fibre source for computer papers.

In Brazil, productivity per hectare, per year, for the Aracruz pulp
mill, is about fifteen times greater than in Scandinavia and northern
North America. Brazil's eucalyptus plantations and land are already
the base for three major pulp mills, with more in construction and
planning. Half the output goes to the export market, and Brazilian
manufacturers anticipate continued growth because, they say, "Eucalyptus is now a desired fibre worldwide." Alberto Fernandez Sagarra,
Director of FICEPA (a Brazilian Forestry organization), says:

> The forestry resources [have] a rapid growing rate when compared
> with the Northern Hemisphere, offer low wood cost, high productiv-
> ity per unit area, great land spaces ready for reforestation, and bio-

mass resources as an energy source. The pulp demand in the international markets along with increasing prices and low labor costs are promoting a leading technology for eucalyptus pulp and paper manufacture as well as more efficient mills. . . . There is the possibility to transform a part of the foreign debt loans to risk capital, and the participation of interested groups in existing and new projects.

Northern Hardwoods

Hardwoods grown in northern regions have also become potential fibre sources. The aspen of the Canadian boreal forest is now susceptible to full-scale logging. Even a decade ago, logging this region was considered uneconomical.

All of the new plantation projects, and the large boreal forests, are attractive to large companies. The investment in plantations yields crops quickly, and in this respect alone is preferable to investments in reforestation of softwood regions. The boreal forests are simply another rich resource provided by nature.

Other Fibre Sources: New, Rediscovered and Recycled

There are some new and rediscovered fibre sources that do not require high capital investments and huge mills. Some countries are beginning to discover their advantages, among which are greater self-sufficiency in pulps and less reliance on foreign investment.

For example, a tropical plant called kenaf has fibre characteristics similar to those of softwoods. It requires about eighteen weeks to reach maturity on flat land in tropical or warm temperate regions. Its disadvantages are that it has a limited harvesting period and cannot be stockpiled for a mill, and it shares with other plants the uncertainties of seasonal production. It would become economical, then, only if mills were located adjacent to both resource and ports, and some of the uncertainties could be eliminated. Australian researchers are engaged in seeking solutions.

While economic calculations made by these researchers assume large mills and export markets, small countries have a different reckoning for the advantages of kenaf. It can be grown in a fraction of the time needed by pines, and provides the cheapest long-fibred component to blend with tropical hardwoods, straw or bagasse pulps.

New fibre sources also include synthetics and sugar bagasse. Synthetics are not yet highly developed sources, and experimentation is still in the early stages, but several mills have been constructed, using them to produce various grades of paper. Sugar bagasse is the raw material for over 100 (relatively small) mills, in Taiwan, Cuba, Colombia, Indonesia, Peru and Mexico, and others are being constructed in India and elsewhere. It has been a problematic source for various technical reasons, but new mechanical pulping techniques promise to make it a major new source of pulp fibre.

The technology for using agricultural residues has an ancient history, but its development was delayed when North American forests began to be logged. Straws from rice and wheat production have again become potential sources of pulp in developing countries. These can be milled without expensive machinery in countries where the raw materials are available.

There has been a rise in the world's non-wood fibre pulping capacity from just under 7 percent in 1970 to over 9 percent in 1990. In 1983, the UN Food and Agriculture Organization (FAO) expected the increase in the proportion of non-wood fibres used in developing countries to double within three years, and to increase substantially in eastern Europe, Italy and Spain. Between 1976 and 1982, their annual rate of increase was 3.8 percent in contrast to an average annual increase of 1.9 percent for wood pulp. According to researchers in India:

> Hardwoods can only be used in large paper mills which require considerable investment and foreign exchange for import of several items of machinery. Agricultural residues like bagasse, rice straw, wheat straw and jute sticks could, however, be used in small plants requiring low capital investment and no foreign exchange as machinery required are at present produced indigenously.

In India, where only 11 percent of the country retains any forest, bamboo has provided a major source of pulping fibres in the recent past. It will no doubt continue to do so, but the available resources are already allocated, and the demand for newsprint and other papers in these countries is growing beyond the capacity of bamboo resources. This provides the impetus for development of straw and other agricultural residues as fibre sources for pulp. China is the leading user of non-wood fibre pulp.

It is not only tropical and underdeveloped countries that are discovering the advantages of non-wood fibres. Mills using straws as raw materials have been constructed in Spain, Italy, Hungary, Bulgaria and Romania. There are proposals to build mills in the United States and western Europe. However, in the advanced industrial countries, the impediment to development of straw resources is high labour costs involved in straw collection and preparation.

Finally, and no longer insignificant in the scale of fibre sources, waste paper and recycled newsprint are becoming major sources. They have provided 40 to 50 percent of Japan's newsprint fibre sources for some time. The FAO estimates a rise in consumption of waste paper from 49 million tons (26 percent of total fibrous raw materials) in 1984 to 84 million tons (34 percent) in 1995.

With new legislation in the United States demanding that newspapers utilize recycled paper, de-inking facilities and new mills based on recycled fibres are coming on stream. These mills need not be located close to forests: they are better situated close to large industrial centres. The developing countries are not the beneficiaries of this, since it is the industrial countries that have potentially copious supplies of waste paper. World markets are not yet well established.

Changes in Industry Structure

Growth by Japan and Other Asian Paper Producers
Japan demonstrated that it was possible to create a huge pulp and paper industry with relatively little domestic fibre, by acquiring supplies elsewhere. The growth of the Japanese paper industry after 1960 relied on the availability of logs, wood chips and raw pulp from nearby Asian countries, New Zealand, Russia and North America, plus recycled newsprint from domestic sources.

It also demonstrated that American companies did not have the monopoly on economic imperialism, and even America itself, rich in resources compared to most other regions of the world, could become a resource reservoir for another country's industry. The major supplier of softwood logs to Japan since the mid-1970s has been the United States, providing about half of the total softwood log imports. The USSR has provided about 35 percent, with New Zealand, Canada, Indonesia, the Philippines, Malaysia, Taiwan and Chile all supplying smaller amounts.

Japanese imports of wood pulps, which are a step up from raw logs and wood chips, nearly doubled between 1973 and 1983. The suppliers were the United States, New Zealand, Canada, Sweden and, by the late 1970s, South Africa and Brazil. These wood pulps are converted into paper and paperboard products sufficient to meet Japan's domestic demand and to permit it to export.

Japan also produces lumber, plywood and veneer, primarily for its domestic market. The major suppliers of hardwood logs for these purposes have traditionally been Malaysia, Indonesia and the Philippines. Export restrictions were imposed during the early 1980s by Indonesia and the Philippines, and since 1983, Malaysia has captured about 73 percent of the Japanese market. Papua–New Guinea and the British Solomon Islands became new sources in Asia, but the major new sources were North America, New Zealand and the USSR.

Until the late 1970s, Japanese companies procured their raw materials elsewhere mainly through merchant activities. But as supplies became scarcer and North American companies began looking for alternative supplies, Japanese companies began to invest elsewhere.

Cenibra, established in the Minas Gerais state of Brazil in 1977, is a joint venture between the state-owned Brazilian mining, minerals and industrial giant, Compania Vale do Rio Doce, and a consortium called Japan–Brazil Pulp and Paper Resources Development (JBP) made up of 18 Japanese paper- makers and administered by the Japanese engineering and sales company, C. Itoh, together with the Overseas Economic Cooperation Fund, a Japanese government agency.

Despite the shaky Brazilian economy and interest rates upward of 700 percent, Cenibra, with cheap wood resources, low costs, and a growing demand for eucalyptus pulp, is expanding its share of export markets. The Japanese companies take half the output of the mills, manufacturing it into high grade papers in Japan. As well, C. Itoh has started expanding sales from this mill to China.

Mitsubishi formed a joint venture with Forestal Colcurra of Chile in 1987, to export chips to Japanese producers, but Japanese investment in Chile lags behind American and New Zealand investments there. There are several joint venture arrangements between Japanese companies and either governments or private groups in Indonesia, providing both pulps and plywood for the Japanese market, which is the major market for Indonesia.

The Japan Paper Association announced in 1988 that it would grow eucalyptus trees in Thailand under a joint venture arrangement with the Thai government. Plans call for the construction of five wood chip plants, and a plantation of 200,000 hectares within five years. The consortium includes Oji Paper.

Japanese companies have also invested in North America. Among the main investment companies is Daishowa, Japan's second largest paper and paperboard manufacturer. This company's strategy is to create joint venture pulp mills with American companies that already have extensive harvesting rights in Canada and the United States. Among the Canadian mills with these arrangements are two at Quesnel, one a joint venture with West Fraser (US), and the other a joint venture that includes Marubini (Japan) and Weldwood (itself owned by Champion International, US). The mills provide a guaranteed market for the logs and chips from sawmills owned by joint venture companies, and the pulp produced in Canada is then shipped to parent companies in Japan for utilization in the manufacture of finished papers and paperboard.

Direct investment in North America increased during the early 1980s, when North American companies were suffering through a sharp recession lasting nearly two years. In addition to greenfield mills planned for the boreal forests of Canada, and the joint venture mills that link Daishowa to American company harvesting rights in British Columbia, Daishowa acquired a James River mill in Washington State and Reed International's North American Paper Group in 1988. The US acquisitions are not intended to supplement Japan's supply of pulp; they are expected to provide the inroad to the American domestic market for newsprint.

It should be noted that unlike the American and European companies, few Japanese firms are reported to have locations in numerous countries. This is partly a function of inadequate reporting. More importantly, it suggests that more Japanese activity is directed primarily toward obtaining raw material, while American activity is directed toward establishing production units elsewhere; and that much of Japanese investment takes form within consortia and intricate joint ventures in developing countries.

Following the Japanese model, South Korean paper manufacturing became established on the same principles. The Korean companies are not yet in the global marketplace, but they have substantially reduced Korea's imports of paper, and increased Korea's fibre imports from elsewhere. Other Asian newly industrializing countries are adopting the same model, and simultaneously attempting to curtail logging in their own territories. The Taiwanese company, Yuen Foong Yu, for example, is one of the applicants to build a pulp mill in northern Alberta.

Total Asian newsprint capacity was expected to grow by some 81,000 tonnes by 1992, with new mills throughout the region, in Korea (Chonju), China and Malaysia, as well as Indonesia and Japan. ASEAN countries became producers and exporters between the mid-1950s and late 1970s.

Indonesia, formerly a net importer, is now exporting some 23,000 tonnes per year, and new capacity is on line. The Indonesian industry anticipates supplanting Japan as the world's top producer of paper and paperboard, drawing on the resources of neighbours and new producing regions.

Tied supply for Asian companies (from mills with joint venture ownership and market contracts) are providing most of the 1.2 million tonnes imported to the region. This includes the joint venture between the Singapore Newspaper Services and CIP Gold River project, the Howe Sound Canfor–Oji joint venture, and the proposed China–Westar project, all in BC.

Fletcher Challenge of New Zealand

Other companies besides Japanese and Asian ones have become major players. The most notable is Fletcher Challenge of New Zealand. The original company began experimenting with exotic pines in the 1960s and burned off old-growth forests in order to plant radiata pine in the 1970s. Starting as a relatively small player in the forest industry, Fletcher Challenge (formally established as such in 1981) acquired several other companies in New Zealand before becoming a global player. It owned Tasman Pulp and Paper, New Zealand Forest Products, and Carter Holt Harvey by the late 1970s.

In 1983 Fletcher Challenge acquired Crown Zellerbach Canada. In 1986, it purchased half of the Papeles Bio-Bio newsprint mill in Chile (later increasing its holdings to 100 percent). In 1987, it bought into BC Forest Products, and subsequently acquired 72 percent of the shares. In 1988, it acquired 50 percent of the holdings in Australia's only newsprint producer, Australian Newsprint Mills.

The move to Canada provided Fletcher Challenge with a large softwood forest reservoir. To make this move, Fletcher Challenge, like Daishowa, took advantage of its relatively fluid financial condition while Canadian and American companies were in straitened circumstances in the early 1980s. Its holdings in BC include 72 percent of all of the former BC Forest Products timber and mills, together with the former Crown Zellerbach timber and mills.

The move to Chile averted competition, since large radiata pine plantations there were already scheduled to reach maturity at the same time as the identical plantations in New Zealand. Both New Zealand and Chile are export-oriented and have small domestic markets, and both are aiming at the Asian markets. The purchase in Chile was made through debt swap financing.

Through Carter Holt Harvey, as well as in its own name, Fletcher Challenge acquired both plantations and other mills. Lumber, plywood and pulp mills are either in construction or planned, and one of these may involve Daishowa as a partner.

In 1988, Fletcher Challenge moved into Brazil, obtaining half the shares of the Papel de Imprensa S.A. (Pisa) mill through debt swap financing. Some of its BC-produced newsprint could be allocated to Brazil so that Fletcher Challenge might optimize capacity in the softwood region while preparing for additional capacity in Brazil. Having supplies in diverse regions puts this company in a market-control position. It also allows it to avoid stoppages and failed shipments due to strikes, political events, natural disasters or other impediments that affect market supplies from single-region companies.

By 1988, Fletcher Challenge had become the largest producer of market pulp, the second largest newsprint producer, the third largest lumber producer and, with all forms of forest production combined, the fourth largest forest company in the world. It owns or has harvesting rights on 3,386,250 hectares (13,074 square miles) of land in Canada, the US, Australia, New Zealand, Chile and Brazil.

However, two years later the Fletcher Challenge empire began to disintegrate as the parent company ran into cash flow problems. The downturn of the pulp market in the early 1990s combined with tougher conditions of operation in BC gave Fletcher Challenge the incentive to sell many of its holdings in BC and in Chile.

Canadian and US Companies

Canadian and US companies (these are virtually indistinguishable) have either been displaced or they have, themselves, internationalized their holdings. Weyerhaeuser, International Paper, Georgia–Pacific and James River have retained their dominant positions by investing in new plantations in warm climates and linking up with capital elsewhere. They have also been active in takeovers of their weaker competitors, several of which have disappeared from the ranks over the past few years. Great Northern Nekoosa (US), ranked sixteenth in 1988, was subsequently taken over by Georgia–Pacific (US).

Abitibi and Bowater entered a joint venture with the Venezuelan government and major Venezuelan publishers (the company has 67 percent of the shares; Venezuela has 18 percent; and the publishers, 15 percent) to build a greenfield newsprint mill near Cuidad Guyana, based on Caribbean pine, with output to be marketed by the Canadian partners. This was put on hold in 1990 because of market conditions. Scott Paper, together with Shell and Citibank, took over the Papeles Sudamerica Nacimiento eucalyptus kraft mill in Chile in 1988.

Noranda linked up with North Broken Hill Peko of Australia (which moved from rank eighty-seven in 1987 to forty-seven in 1988) in a new company, Wesley Vale, to produce plantation eucalyptus-based pulp in Tasmania. Although rebuffed by a strong environmentalist lobby on its first round in 1988, the joint venture has taken on a second life now that the Australian and Tasmanian governments have concluded that the proposed plant will be safe for the marine environment. Noranda has sold its MacMillan Bloedel holdings, so that its parent company, Brascan, could obtain cash funds. Noranda itself still owns Norwood and other smaller pulp producers in Canada. Kimberley–Clark established a joint venture with private groups in Sumatra, Indonesia.

These companies still have extensive forest harvesting rights in softwood regions of North America, and several have private lands in the United States. This base continues to be their strong suit, but their futures lie in the new plantations, either off-shore or in the southern United States. In the United States, a shift from the north-western to southern regimes is clearly taking place.

In Canada, there is growth based on the remaining softwood forests and in the new hardwood forests, but most of this growth is occasioned by foreign direct investment driven by a search for raw materials; there is no real growth in manufacturing capacity beyond basic pulp mills.

Nordic Countries

As noted above, the Nordic countries have not deforested their own lands. Reforestation and afforestation projects at the turn of the century in Sweden, Norway and Finland have continued to provide them with second-growth fibre supplies. Until the 1960s, they were peripheral staples producers, similar to Canada. However, with the unification of Europe underway, Swedish and Finnish companies began to diversify their holdings in Europe.

The Finnish company Kymmene and the Swedish company Stora have established large-scale production units for printing and writing papers within the European Economic Community. Kymmene has attempted to obtain fibre supplies from the southern United States to supplement Finnish sources. Stora has obtained eucalyptus plantations in Portugal, and has expanded as well to Brazil and North America. It is a major shareholder in the Aracruz mill in Brazil.

The Industry Now

The organization of forestry in the 1960s, with Nordic countries providing raw material for European paper-makers and Canada providing the raw material for United States paper-makers, is still in place, as data on fibres above indicate, but no longer does it have the same global context. The growth of pulp and paper producers around the world has fundamentally altered the structure of the industry. Swedish, Finnish, American and Canadian companies are internationalizing their operations, buying into established mills or creating joint-venture greenfield mills wherever they can obtain new fibre resources.

The established companies and consuming nations have tried several options to procure new supplies. Originally, they did what resource-seeking buyers have always done—produced the raw material (i.e., logs) elsewhere, once technologies were in place to utilize them. Most countries have tried to resist this exploitation, though log export bans have not always been either possible or enforceable.

The second option is to build sawmills and pulp mills near the potential resource base and, in the case of the new tree species, invest in plantations. These new mills are typically joint ventures with governments or with private capital in the developing regions, with tied-market conditions whereby the larger part of the output is sold to the parent paper-producer in the home country. Occasionally, the joint venture mills become large enough, or the plantation base appears sufficiently promising, that the new mills develop independent markets. Rarely do they produce more than crude pulp.

The huge mills built by these international companies are capital intensive. They rely on cheap wood raw materials, but the risk to capital in a notoriously cyclic pulp market is considerable. Salonen and Niku, two analysts favourable to large concentrations of capital, note:

> To emphasize the importance of capital costs, we may say that only the interest charges for such a huge investment during the two to three year period would equal the total investment costs of a competitive newsprint machine to be built in Scandinavia. The main message here is that the industry needs much more of a risk-taking capacity than we have been used to so far. Since the markets are still growing, we will see larger and larger international groups being formed.

Mergers and joint ventures became more frequent in the mid- to late 1980s than at earlier times, and more of these crossed continents than ever before. In addition to acquisitions by established companies, new mills are being constructed which combine these pools of capital with new groupings within developing regions, and with state equity in most countries.

The net result is a greater concentration of total holdings in the established softwood forest regions, and an expansion of holdings by established companies elsewhere; but, overall, a wider spectrum of capital sources are included in the joint ventures, and a much greater participation level for groups that had no holdings in forestry just two decades ago.

At the present time, eight companies control about 40 percent of the US paper market. This is more concentrated than was the case a decade ago, but it is not as concentrated as most manufacturing industries, and with the new participants elsewhere, new fibre sources and greater self-sufficiency in developing countries, the industry is not moving toward global concentration yet: it is restructuring and stretching, but the overall context has fundamentally altered traditional positions.

What are the implications of this restructuring? Some are obvious: Asia is becoming self-sufficient in wood products, sourcing raw materials in the north and in the south, and not dependent on any one region for its supplies. As well, its machinery companies have linked construction of supplying mills with their own outputs, cutting out their competition in the process. Much of Latin America is now self-sufficient, though without independent manufacturing linkages.

The Nordic companies have recognized the trend and followed suit, aiming now at a larger share of the European market with raw materials from the south and paper mills in western Europe. The United States, with its southern plantations, is also becoming increasingly self-sufficient; in fact, it will likely begin to generate a surplus as its global markets continue to disappear.

Canada, cursed by the lush gifts of nature, has failed to diversify. It sells its boreal forest as it sold its Douglas fir—for the quickest profit. Loggers and environmentalists fight on, seemingly oblivious to the global change in the market value of the trees.

Succeeding in establishing new forestry industries elsewhere, some companies are withdrawing altogether from northern forest regions. Many of the companies that remain or continue to invest in these regions are resource suppliers more than manufacturers; they are shipping quantities of wood in the form of logs and wood chips, or at the most, raw pulp, to be used as inputs for manufacturing companies elsewhere. New mills are capital intensive both in North America and northern Europe; old mills are being phased out.

Impact on BC and Northern Regions

Throughout the world, tropical and temperate zone rain forests are being rapidly depleted. Their wildlife and oxygen-generating capacities are disappearing, and human societies dependent on them for sustenance are suffering either cultural (and sometimes physical) death, or dislocation.

There are also social impacts in the northern softwood regions, where employment has steadily declined. Mechanization and automation have allowed companies to increase production and productivity per worker with declining labour demand. Lonnstedt estimates a job loss in Swedish forestry of more than 3,000 per annum after 1980.

Data for British Columbia alone suggest even higher rates in North America's traditional forest regions. Statistics Canada shows an overall decline in employment between 1979 and 1986 from 96,841 to 74,306. Community decline attends mill modernization. It is also the consequence of changing markets: Oregon and Washington have suffered continuing and dramatic decline in community stability and overall employment with the rise in log exports and decline in manufacturing over the 1980s. The exodus of companies from single-employer towns is the death knell for numerous old communities along the northern Pacific coast.

BC's Future

These arguments are all by the way of situating the dilemma we experience in British Columbia within a larger context. The implied argument throughout is that much of the debate over forestry in BC is based on the false assumption that the industry will continue indefinitely as the major economic activity of the region.

British Columbia has long enjoyed the riches derived from a lush temperate rain forest. Its softwoods have commanded high prices on world markets where only the nearby northwestern states and the Scandinavians could seriously compete. Within the past decade, however, the entire industry has changed.

Tropical and sub-tropical regions can now produce fibres suitable for wood and paper products that were formerly produced only from northern softwood forests. As well, non-wood fibre sources are becoming significant in production of certain paper products. New markets controlled by different pools of capital have emerged, and global corporations are now established. The ownership structure of the industry has undergone dramatic changes. Environmental concerns are beginning to affect the industry. In BC and the northwestern United States, the softwood forest has suffered depletion, and in the new global context there are few economic incentives to replant it.

Canadian and US companies overinvested in lumber as in pulp mills over the past two decades, and the landscape on both sides of the border is dotted with enormous sawmills. At the moment, and for some time to come, these mills collectively produce much more lumber than the market can buy, so that increasing production is providing lower dollar returns. We are now cutting irreplaceable old-growth timber to sell at prices even the producers say do not meet their costs.

The market for standard-sized lumber is declining. The age of ranch-style housing in North American suburbs is passing as fuel costs increase and land becomes scarcer. An increasing share of accommodation consists of multi-family buildings, and these are not built of wood. Alternative markets want quite different products.

But there will be a market for softwoods for interior construction and specialty items. With the shortage induced by past failures to restock our forests, the remaining lumber-grade timber will be extremely valuable.

In time, then, the investment companies will have a resource to sell, and new, small, automated mills will be established. Meanwhile, the tariff wars have taken their toll on companies with too much capital tied into the old mills. They will also take their toll on the International Woodworkers of America, as employment and membership decline in an aging industry.

No BC or Canadian government could reverse these changes, but a government could alter the impacts considerably. Instead of waiting for outside investors and imported machinery for new mills, a BC government could take the initiative to create an industry under BC ownership and management. It is within government powers to restructure the stumpage and tenure system so that investors would move out sooner rather than later and reallocate tenures to communities and unions under a contract arrangement so that they manage the forests.

The machinery could be manufactured in BC, and woodworkers could be retrained for the jobs both in machinery and manufacturing plants and in the new mills through an infusion of funds into vocational training throughout the province.

These possibilities are well within the powers of a provincial government, and the time for acting on them is right now, while the major investors are looking to southern climates for their profits.

Seeking Solid Ground

In the Environmental Assessment Quagmire

by Julian Dunster

CANADA, WITH THE SECOND LARGEST MASS of any country, has many conflicts about the use of its abundant forest lands. On one hand, an inexorable industrial demand for wood is nimbly fostered by bureaucrats, industrialists and professional foresters, many of whom still regard forests as little more than two-by-fours and pulp waiting for their attention, and virtually all other uses of our forests as a waste of good wood. On the other hand, the Canadian populace is realizing that the magnitude of our forests is more than matched by the enormity of this industrial demand, and that widespread environmental and social impacts are likely to be the result.

Small wonder then, that a broad spectrum of society, both in Canada and around the world, believe that not only should Canadian forest practices be improved, but also that less of the forest should be designated for logging—the most pervasive single use activity, and ecologically, surely the most drastic disturbance visited upon the Canadian landscape in a short period of time.

It has often been suggested that one way of achieving the improvements thought necessary would be to undertake an environmental assessment (EA) of various aspects of the forest (timber?) industry, government policies, and all associated aspects.

Perceptions about the merits of environmental assessments vary. Many people in the timber industry seem to view EA as a waste of time and money, but a necessary evil on the route to gaining project or process approvals. For those in the environmental activist movement,

EA is often seen as a means of proving that current practices or processes are automatically wrong, and of assuring that "better" practices replace them. Government officials seem to regard EA in several ways depending on their subconscious mindsets. One group seems to regard EA as a useful means of providing career jobs for planners, computer analysts, text writers and researchers.

The second group seems to follow the industrial attitude that EA is a necessary but unwanted evil, placed in the path of progress to placate politicians and bleeding hearts. A third, though extremely minor faction in most cases, actually believe in the purity of well founded scientific research as a credible means of providing definitive answers. This latter group tends to naively believe that political interventions in the decision making process will not distort scientific "facts," and that "objective" assessments will overcome "subjective" debates—a concept nicely debunked by D.A. Bella in his seminal article "Ethics and the Credibility of Applied Science."

What Is Environmental Assessment?

In practice, an environmental assessment typically involves many people and an array of attitudes encompassing all of the above. However, it is often unclear what advocates of environmental assessment really want, and how they believe it will help to resolve the many difficult land use and land management issues now confronting us. Nor is it always clear why other groups ceaselessly argue against EA. Cynically, one might interpret a reluctance to undertake an EA as a tacit acknowledgement that the answers derived will be unpalatable, or worse still perhaps, that the answers might raise more awkward questions (almost inevitable) and reveal the true extent of our collective ignorance. The truth is that we would all benefit from better knowledge and understanding about how we treat our forests and what the outcomes of the treatments will or may lead to.

Environmental assessment is really little more than a well structured planning process that permits a systematic review of planning proposals before they are put into action. The aims of EA are to try to assess what will happen—where, when, how, and why—and then to see if the desired goals might be achieved in different ways, so as to maximize the benefits for people and the environment and minimize the losses, disturbances and damage.

A fundamental tenet of all environmental assessments is that the onus to prove benign outcomes lies with the proponent of the project. That is, the people proposing a new project or policy are responsible for clearly demonstrating its need, its sound design, and that the inevitable environmental and social costs are clearly outweighed by the benefits. This responsibility is all too often abrogated and usurped for devious political ends that have little if anything to do with impact assessment. Advocates and antagonists have often successfully manipulated assessment processes in many ways, subverting the original spirit and intent of EA under the guise of political expediency, cost effectiveness, and short-term gain for a few at the long-term expense of many. Good examples of clever manipulation are seen in Ontario, where the Ministry of Natural Resources first assessed the construction of a logging road into the now famous Red Squirrel area, and then undertook to carry out a separate assessment of the actual logging proposed. But it was argued in court that the latter was so clearly a result of the former that the two activities could not be assessed independently. (The Crown won on a technicality inherited from a blanket exemption granted several years earlier.) Similar abuses of EA are seen in Manitoba's sale of a pulp mill, in which the new owner undertook to assess pulp mill expansions and the creation of a new pulp mill, without first considering if a supply of wood existed, and what the effects of cutting huge areas of forest would be. Quebec also proposed to assess road construction independently of the dam construction that these roads would service.

In all of these cases, the proponent tried to break a large project into smaller assessments. Superficially, this may seem logical, but of course, once the first assessment to construct a road is approved, how likely is it that the rest of the project can be stopped, even if the need is then found to be unjustified? Moreover, several reviews of EA processes around the world reveal that EA documents often contain a combination of elaborately stated theoretical possibilities—frequently based on tenuous assumptions—and dubiously predicted results. Evaluations show that many of these EA products have yielded a strange array of anticipated—but unrealized—benefits, a problem that besets other forms of planning as well.

Unfortunately, many people view EA as a panacea for almost all problems, and the process has gained an almost mystical aura of credibility in many people's imaginations. Still, all is not entirely lost because, despite the abuses, a well thought out and carefully implemented environmental assessment does offer the opportunity for a systematic review of planning processes, potential environmental effects, and potential changes to the environments likely to be affected if a process is implemented. It also forces some accountability onto proponents, and allows varying degrees of public scrutiny which might not otherwise be available. Moreover, the advent of legislated assessment requirements and court scrutiny of the processes and products of EA has spawned some excellent research in forestry. This is especially noticeable in the United States where more rigorous legislative requirements have forced some excellent research into wildlife and forest habitat needs for a range of plant and animal species, and it is one of the few evolving benefits in Ontario's Timber Management EA.

Forests pose an interesting challenge for environmental assessment because they encompass a very complex biological system, exhibiting huge variations in the functional components present, and the manner in which any one component might be affected by changes elsewhere. The crux of the problem in assessing forests is to know which parts of the whole must be assessed, and in what detail, in order to have a defensible basis for good decisions.

The US Approach

The application of environmental assessment procedures originated in the United States, with the National Environmental Policy Act (NEPA), which came into effect in 1970. NEPA mandated preparation of an assessment by federal agencies, forcing them to prepare a report outlining the effects of proposed actions and the available alternatives, and to submit the report to general review and debate.

Not surprisingly, early attempts were crude, since they had little or no previous work to guide them. Over time, an enormous number of court challenges have served to mould and better define what is or is not acceptable as an assessment, and several major legislative initiatives have been introduced to rationalize the assessment process while still meeting the spirit and intent of EA.

Despite these new legislative initiatives, many conflicts still occurred. As a means of trying to reduce the number of conflicts and rationalize the forest planning process further, new legislation was introduced in 1976, under the National Forest Management Act (NFMA).

This new legislation saw much more detailed forest planning as the primary solution to conflict, and a new planning process was introduced that mandated each National Forest to prepare a Land Resource Management Plan (LRMP), based on a rational, comprehensive model of planning.

In principle, the requirements of the NFMA were designed to serve the need for comprehensive environmental assessment (thus meeting NEPA requirements), while at the same time providing the better data needed for an explicitly, and publicly accountable, decision-making process. In practice the Land Resource plans are voluminous and complicated documents that require a sound technical knowledge to understand. To some extent, this complexity results from the use of computer modelling, the inputs and outputs of which are typically not simple to follow. This creates some difficulties for public and agency understanding of how results and alternatives have been derived.

Despite the apparent sophistication and versatility of the computer models now being used, the outputs of many of the EAs still have shortcomings, often centred around institutional and individual biases, political agendas and unrealistic expectations. In addition, there is a sense that the more detailed planning efforts have not achieved their potential to actually modify practices on the ground. The main failing, it appears, is not one of process so much as the pervasive corporate ideology of timber primacy, which has always guided the US Forest Service, and, according to its critics, still does. In effect, no matter what planning and management process is used, the Forest Service remains heavily biased toward commodity production (lumber) with all other aspects inherent in forest management being subservient to this primary goal.

But the plethora of plans, appeals, revisions and new plans has also had benefits. These originate primarily in the court mandates given to the US Forest Service, which forced it to undertake the technical research needed to examine the environmental effects of harvesting in forest ecosystems. Much remains to be learned; still, the breadth of technical research work, and the depth of funding commitment to continue it, is much stronger than in Canada.

The Ontario Approach

The province of Ontario has used its Environmental Assessment Act to force the Ministry of Natural Resources (MNR) into preparing an environmental assessment of forest related activities. Initially, the MNR proposed an assessment of forest management, but later scaled this down to timber management to more accurately reflect their institutional preferences. Ontario has adopted a process called Class Environmental Assessments, arguing that individual assessments of repetitious activities are a waste of time when these can all be classified into "classes" of activity.

The Ontario process, which I reviewed in detail with Bob Gibson in 1989, is something of an aberration in the annals of EA. The MNR made five attempts to submit an assessment document that satisfied the government review team, the last one being finally accepted in 1987. In the meantime, the MNR quietly went ahead in 1983 and adopted its own proposals as policy and has been using them ever since. Perhaps even more bizarre, the government initiated a public review of the MNR's proposed assessment in 1988, a quasi-judicial hearing that was still underway in 1992, almost ten years after the MNR went ahead anyway.

While the Ontario process also advocates a rational planning process, along the lines of the US model, it is far less clear how the process actually assesses anything. The process continually refers to a set of manuals and guidelines that are supposed to help in the implementation of well integrated forest management at the local level. But it is unclear how guidelines can be enforced if they are not mandated. Similarly, the MNR adopted timber management, not forest management, as its basis for assessment.

While this realistically reflects the institutional ideology, it neglects the vital fact that timber management is but one component of the whole, namely forest management. In so doing, the MNR has advocated a process of non-integrated assessments, and the real problems of integrating activities in the broader forest landscape remain neglected. It remains to be seen what all of this expensive testimony will achieve, but already several major policy reviews and administrative changes have been initiated.

The hearing process, tedious and expensive though it has been, has, like its US counterpart, produced some incidental benefit. Not the least of this has been the exposure of the many weaknesses inherent in the current system of forest and timber planning. The hearings have served the public and government well, if only to reveal the true extent of ignorance about what is actually happening in the forests. Additionally, the hearings have highlighted the need for data collection that is province-wide and in compatible formats, and for consolidation of the many studies and research initiatives undertaken already or now underway.

In contrast to the US assessment process, the Ontario process only looks at timber aspects (although the hearings have forced much broader debate), omits policy assessments as a means of guiding planning initiatives and has a dubious legislative basis for enforcement.

Hopefully, the Ontario government and the MNR will use these revelations to gain strong political support for the much needed improvements that must be implemented if a usable assessment is to be achieved. But, the public hearing process is taking so long to complete that, conceivably, slow progress seems inevitable, and other jurisdictions have taken Ontario's approach as a warning to avoid environmental assessment of forestry at all costs.

The Potential of Environmental Assessment in BC Forestry

By now it should be clear that environmental assessment is not the simple panacea that so many have hoped for. The application of EA to forestry proposals has been and continues to be fraught with a complex set of technical, institutional and socio-political problems. Conceptually, the application of EA principles and techniques to forest planning and management ought to be feasible. Moreover, the expectation that EA will lead to less serious environmental consequences, as a result of careful scrutiny, more thoughtful designs, and process improvements, is not unreasonable.

Despite this, experience with EA and forestry has often been disappointing. Expectations have not always been met, especially when major policy issues remain unanswered, and many of the crucial cause and effect linkages between timber management activities and environmental quality objectives remain largely unknown, untested and in many cases, simply not even considered.

Environmental assessment is merely a process that can facilitate decision-making. What we call the forest planning and management process, be it EA, integrated planning or sustainable forestry, really does not matter as much as the visible results on the ground, and the degree to which these results match or do not match societal expectations and ecosystem abilities. Ultimately, the crucial need is for technically sound and achievable plans and management actions on the ground; only then can we actually see if the outcomes are beneficial and sustainable.

Moreover, the application of sophisticated forest planning and management processes pose difficult challenges, because the sheer extent and complexity of forest ecosystems defies any simple process by which all actions can be dictated. The management of these forests for multiple uses requires many different planning and management techniques, drawn from several disciplines. Given so much societal and ecosystem complexity, it may be unrealistic to hope that there will ever be just one planning and management process to cover all permutations, no matter what it is called. There are also inherent structural management difficulties. In Ontario, and to a great extent in the United States, the EA process requires the detailed assessment of planning and management options to be undertaken at the local forest management unit level. This has many attractions, especially in the ability to utilize local knowledge and understanding effectively. But, to expect locally based efforts to tackle and resolve the broader policy questions and conflicts may be unrealistic if there is no overall guidance at a provincial level. Lower level planning initiatives need a context for action, and, at the same time, they provide the necessary field experience to guide amendments and refinements of higher level decisions.

Priority Issues

Most contemporary forest conflicts centre around the issue of land allocation (purposes) and the way in which management is undertaken (practices). These two broad policy questions need addressing in most Canadian forest planning and management debates. Both questions form a vital and very fundamental starting point for any assessment issues at the local level, and must be addressed effectively at the broad policy level if local efforts are to function well.

Whatever the process is called—impact assessment, a comprehensive land use strategy, or strategic land use planning (as was attempted in Ontario)—there is an urgent need to actually reconcile the competing interests which are placing the land base under increasing pressure. Aboriginal land claims, demands for parks, wilderness areas, dam impoundments, and ecological reserves, in addition to the need to define a more permanent land base for growing wood fibre, all compete. It is not physically possible to manage for all of these uses at the same time and place; many of them are mutually exclusive options and we have to choose one or the other.

Lessons So Far

The first lesson is that the application of EA to forest planning and management has potential—but, the way in which the EA process is applied is often far more important than the process itself. EA evolved as a means of elaborating uncertainty, in an effort to enhance the decision-making process. As a result, EA practitioners can expect to find more complexity as the elaboration process unfolds. This complexity, some of which will be based on scientific research results, is still used subjectively in the final decision-making processes. Consequently, an equitable EA process acknowledges this and attempts to foster a workable solution that is acceptable to all the stakeholders while still meeting the broad objectives of better environmental planning and management.

Wondolleck provides an excellent review of these problems, and suggests that there are five essential steps for successful decision-making:

1) Building trust
2) Building understanding
3) Incorporating conflicting values
4) Providing opportunities for joint fact-finding
5) Encouraging cooperation and collaboration.

The key to these steps is the way in which participants in the EA process can present their concerns themselves and not have them assimilated by professional land managers. The attitudes of the participants involved in any process are often more important than the details and "facts" involved. Resource managers must recognize this and learn to deal with it.

The second lesson is that there is no one simple process that will address all the problems all the time. The philosophy of adaptive management—expect the unexpected—should remain high on the agenda, and any EA processes used must retain some flexibility to respond swiftly to changing circumstances. That being said, this flexibility must not be abused so as to permit constant avoidance of what are undoubtedly very difficult problems.

A third lesson is that the need for better data is of paramount importance. Many of the cause and effect linkages resulting from timber management remain poorly understood at best; some are little more than speculative guesswork. Meeting this need for better data requires a long-term political commitment to funding technical research. An important caveat is that data must not only be credible, but also be made widely available. In the Canadian context, most EA and forest work occurs on Crown land, and regardless of whether the land is managed by the Crown or by the industry, there appears to be no good reason why publicly funded research should result in privately held results.

The fourth lesson is that the use of EA brings with it the need for a commitment of staff time, funding, and a political commitment to follow the process through fairly. EA is a means to enhance decision-making, but it has to be accepted that in some cases, better information, more credible data, and closer examination of projects, processes, and activities, and all the various alternatives available for these, should lead to major shifts in planning and management. This challenge to established vested interests will create friction, but this should not deter practitioners from improving environmental planning through the diligent use of EA (or any other similar processes).

Finally, EA is not the only means of achieving enhanced decisions. Other processes, involving essentially the same ideals but without EA's entrenched aura, can be used just as effectively. For example, the special area management processes, known as SAM, and developed in the United States, have been utilized very effectively as a means of implementing innovative solutions to resolve complex resource management conflicts. There is no reason why similar efforts cannot be applied in a forest planning and management context. Political commitment, dedicated people, imagination and publicly credible processes (a novel idea perhaps) will go a long way.

Despite all the many difficulties still encountered in the application of EA to forest planning and management, the concepts of EA have the potential to refine planning and management processes, thus enhancing our knowledge and understanding of how to better manage forested lands for sustainable and multiple outputs. The application of EA is still far from perfect, but it has come a long way since the early days following NEPA and, given the chance, can go much further yet in fostering better environmental planning.

Sources and Recommended Readings

Beanlands, G.E., and P.W. Duinker, 1983. *An Ecological Framework for Environmental Impact Assessment in Canada.* Halifax: Institute For Resource and Environmental Studies.

Bella, D.A. 1992. "Ethics and Credibility of Applied Science." In: Reeves, G.H., D.L. Bottom, and M.H. Brookes. *Ethical Questions for Resource Managers.* General Technical Report PNW-GTR-288. Portland: USDA, Pacific Northwest Research Station.

Dunster, J.A. 1992. "Assessing the Sustainability of Canadian Forest Management: Progress or Procrastination?" *Environmental Impact Assessment Review* 12(1/2):67-84.

—1991. "The Use of Environmental Impact Assessment in Forest Planning and Management." In: *Forest Resources Commission—Background Papers.* Vol 1. Victoria.

Dunster, J.A., and R.B. Gibson, 1989. *Forestry and Assessment: Development of the Class Environmental Assessment for Timber Management in Ontario.* Toronto: Canadian Institute for Environmental Law and Policy.

Gibson, R.B. 1988. *Lessons of a Legislated Process: Twelve Years of Experience with Ontario's Environmental Assessment Act.* Paper presented at the Annual meeting of the International Association for Impact Assessment, Brisbane, Australia, July 5-9, 1988.

Hyman, E.L., and B. Stiftel, *Combining Facts and Values in Environmental Assessment: Theories and Techniques.* Boulder: Westview Press.

Tripp, D., A. Nixon, and R. Dunlop, 1992. *The Application and Effectiveness of the Coastal Fisheries Forestry Guidelines in Selected Cut Blocks on Vancouver Island.* Nanaimo: D. Tripp Biological Consultants Ltd.

Wathern, P. Ed. 1988. *Environmental Impact Assessment: Theory and Practice.* London: Routledge.

Wondolleck, J.M. 1988. *Public Lands Conflict and Resolution: Managing National Forest Disputes.* New York: Plenum Press.

Forest Practices

Putting Wholistic Forest Use into Practice

by Herb Hammond

BEFORE WE TALK ABOUT CHANGING OR IMPROVING forest practices, it seems obvious that the first thing we have to do is ask a question: What do we mean by "forest practices?" One way to answer this is with a very simple definition. Forest practices are human activities—any human activities—that affect the forest. I want to come back to this simple definition. First, however, we need to acknowledge that there's another answer to the question.

In the last hundred years in British Columbia, industrial and government timber managers have used the phrase forest practices to mean large-scale industrial timber production activities, such as road construction, clearcutting, slashburning, brush control, establishing tree plantations and pesticide use. Although this meaning of forest practices is extremely narrow, it is also very widely accepted by those who equate "forests" with "timber" and "forestry" with "timber production." For example, the 1991 BC Forest Resources Commission Report includes a major section entitled "Forest Practices." In it, the Commission recommends that a "single, all-encompassing code of forest practices" is required to govern "all the elements of good forest management, such as:

— the choice of clearcutting or selection logging as a harvesting system . . .
— the size, distribution and timing of clearcuts.
— approaches to be taken during harvesting to consider a wide range of other values . . . "

The three items listed above comprise the Commission's complete list of examples of "elements of good forest management." The Commission's other recommendations under the heading of "forest practices" concern timber cutting systems and road construction.

What is significant here is that the BC Forest Resources Commission obviously interprets the word "forest" to mean "timber" and the phrase "forest practices" to mean "large-scale industrial timber production activities." Although passing attention is given to "other values," the basic assumption is that, in order to do a better job of managing forests, we need to get better at cutting timber. The Commission's advice that we must "consider a wide range of values" during timber cutting makes the implicit assumption that timber cutting is the central use for the forests of British Columbia and all other values (such as biological diversity, animals, plants, water, soil, ecosystem integrity, wilderness, recreation, tourism, ranching and trapping) will fit in somehow.

Put in "soft" terms, this way of defining forest practices represents a narrow, human-centered approach to forest use. Put bluntly, this is forest exploitation for short-term timber benefits.

The Basis of Conventional Forest Practices

In planning forest use and defining forest practices, all of us have to keep learning all the time, applying principles derived from both the newest scientific thinking about ecosystems and the oldest cultural traditions about our part in nature. However, our current forest practices don't really have a foundation in either science or ancient cultural ways. So how did we get to this point? It's important to know something of the history for two reasons.

First, we need to be sure that we are not fooling ourselves when we set out to make changes. We can't just redefine terms such as biological diversity so that the new definition supports bankrupt approaches such as clearcutting. We must change our way of thinking.

The second reason is that current practices are still current. We need to have knowledge and perspective because most government and industrial timber managers—the individuals who hold decision-making power about forest use in BC—still subscribe to a narrow, human-centered, timber-biased view of forest practices.

Historical Roots

Silvicultural techniques, such as clearcutting, tree plantations and brush control, were brought to the United States and Canada from Germany about a century ago. Although these techniques are sometimes called "traditional" European forest practices, the techniques are in fact only about 200 years old and were developed initially as an attempt to rehabilitate the severely degraded forests of central Europe.

According to German forester and historian Richard Plochmann, by the early 1800s central European forests were "overcut, overgrazed, overraked, and overbrowsed . . . the forests were exhausted and degraded." More than two-thirds of the indigenous forests of central Europe had been cleared for farmland and the forests that remained had been heavily exploited for timber. Plochmann explained in his 1992 *Journal of Forestry* article, "Not one acre of forest was left untouched . . . on 10,000 acres of one forest district, no tree could be found strong enough to hang a forester on."

The methods selected to "rehabilitate" these degraded forests were strongly influenced by the advent of scientific forestry, by the lure of market economics, and by the result of an intense debate regarding the earthly role of human beings. On one side were those who believed that humans were meant to be part of the forest and part of the earth, and on the other side were those who believed humans were meant to dominate nature. We know, by now, which side won— and we are beginning to understand the price of that victory.

Rather than restore degraded forests with native species, European foresters adopted silvicultural practices that included planting selected species in order to bring the highest possible economic return for timber in the future.

In the very broadest terms, the philosophy of domination over forests led to forest practices intended to make the forest "more productive" and to "normalize" the seemingly random and unpredictable events of nature so that controlled and predictable processes and products would result.

Legislated Timber Bias

Transported to North America and enhanced by 20th century technology, European-style silviculture has resulted in a set of conventional forest practices that do not make a distinction between "the forest" and "timber." Forest management is equated with timber management. Conventional foresters, quite literally, cannot see the forest for the trees. To a conventional Canadian forester, if trees are growing on a site, it's a forest. If it's a forest, it's a potential timber supply. If the timber is removed and more trees are planted . . . it's a forest again.

This mode of thinking is known as "timber bias." All agencies empowered by the Forest Act of BC—the Ministry of Forests, the Forest Service and the timber industry—exhibit it. With the exception of a few provisions for "managing wilderness," the Forest Act of 1978, which controls the management and use of public forests in BC, is wholly concerned with three activities: first, distributing timber-cutting rights; second, guaranteeing that a certain volume of timber (called the allowable annual cut or AAC) is actually logged and removed from the forests each year; and third, requiring the regeneration of commercially valuable timber species on logged areas.

Evidence of timber bias is abundant. Consider the following examples:

— The only forest value or activity that has strong legal definition and legal protection in BC is timber extraction. The Forest Act provides little or no legal protection for animals, non-commercial plants, water supply or water quality, soil, jobs, cultural values, recreational values, tourism, landscape ecology, stand ecology, or, most importantly, biological diversity. (A more appropriate name for this piece of legislation would be the "Timber Act.")

— No planning for the use of any public forest in BC takes place until some entity—usually a corporation—wishes to remove timber from the site.

— About 90 percent of the timber extraction in public forests in BC is clearcutting—a form of deforestation. The individuals who plan and approve this deforestation are called "foresters."

— Data and information about public forests collected by the Ministry of Forests relates primarily to timber extraction. The Ministry itself collects and maintains very little data about water, soil, climate, animal and plant populations and their habitat needs, wilderness values, recreation and tourism values, ranching values, or trapping values. Given the data it collects and maintains, all that the Ministry of Forests can confidently "plan" or "manage" is short-term timber extraction, not forest use. (A more appropriate name for this governmental agency would be the "Ministry of Timber.")

Forest Politics

Current forest practices in BC are firmly entrenched in the politics of timber, and these practices will not change until forest politics change. Forest politics will not change until the education and political systems provide a more balanced view of the forest and the place of human culture in the forest. As we have seen, legally-sanctioned public "forest use" in BC is essentially synonymous with industrial timber extraction—as fast and efficiently as can be justified by political and industrial institutions. This limited and distorted concept of forest practices is the result of industrial control of the public forest, achieved by means of carefully designed forest tenures. Once a forest is included in a conventional timber industry tenure, all uses except timber extraction become tentative, at the discretion of short-term timber cutting objectives.

Once the forest is clearcut, most non-timber uses and values, such as wilderness, tourism, fisheries, wildlife habitat, soil and slope integrity, or water quality, are usually degraded or destroyed for one or more human generations. Conventional forest practices contribute significantly to our largest problem: loss of biodiversity.

The Ministry of Forests speaks openly, in its own publications, of its "partnership" with industry, of what it calls a "co-operative relationship with its chief clients—a relationship that has become, in effect, joint stewardship of the forest resource." Even though the forests in BC are 95 per cent "publicly owned" (if anyone can "own" a forest), this concept of ownership does not include control. It's hard not to see a prison metaphor here. Prisoners "own" their bodies, but are denied control over how they use them. The public of BC ostensibly "owns" the vast public forests in BC but is denied control over what goes on there.

The Ministry of Forests, the timber industry, and many foresters maintain that the public does have control over forest use and forest planning, pointing to numerous public involvement programs. However, according to official Ministry policy, "decision-making status" belongs to Ministry personnel and to "those under legal contractual obligations as specified in the statutes of British Columbia." Groups and individuals "without legal contractual relationships with the Ministry" have "advisory status" only.

Excluded from "decision-making status" are many groups whose interest in the forest is wholly legitimate (but not dominated by short-term economics)—groups such as Indigenous people, rural water users, ecotourism operators, ranchers, trappers, small businesses in rural forest-based communities, wilderness users, artists, scientists, educators and future generations of Canadians. All of us, in fact, have a legitimate interest in the forest, as do all of the non-human parts of the forest who cannot speak for themselves. A value need not be profitable to be legitimate.

Our most important interest in the forest, while not dollar oriented, is purely self-serving: a natural, biologically diverse forest is essential to our survival.

Current public involvement is not meaningful involvement. By the time the public is "invited" to join the planning process (in its advisory-only status), the most significant decisions have already been made. Tenures have been assigned and the AAC has been determined and approved. Under the terms of current legislation, public involvement is little more than public pacification.

The Failure of Conventional Forest Practices

Public Discontent

The public is increasingly not being satisfied or pacified. Not only is there a call for different timber extraction practices—ecologically responsible selection systems as opposed to clearcutting, for example—there is a growing call for a complete overhaul of the contractual arrangements that give control of public forests and forest practices to the strongest special interest group—the timber industry.

In response to public pressure, the timber industry and government timber managers often emphasize the importance of being "objective" or "impartial." As the BC Forest Resources Commission

Report of 1991 tells us, "It is important to separate the fact from the emotion in this debate." Industry and government representatives speak as if "objectivity" and "impartiality" were something real.

Unfortunately, they are not. Objectivity and impartiality have always been an illusion. They used to appear real, in regard to the forest, because we used to have low populations, low levels of consumption, low levels of pollution and economic pressure and huge untouched expanses of natural forests. So it was easy to be objective and impartial. It was easy to say that people who want to cut timber can do that in the valley over there, and people who want to hike in old growth forests and watch birds can do that in the valley over here, and you can all stay out of each other's way. It was simple and easy to figure out—it was simple to be objective.

But now we have a large population, high levels of consumption, and high levels of pollution and economic pressures. So much timber has already been cut that we can't stay out of each others' way any more. In many parts of the province of BC, logging is now taking place in the same watersheds that provide cities, towns and rural residents with domestic and agricultural water supplies. It is finally clear that what we really have is one forest to serve many values. And suddenly you can't find an objective or impartial person anywhere.

We're left, not with the need to "be objective," but with the need to choose which ethic we want to follow. This is not a matter of finding facts—everyone will interpret the same set of information according to their own values. Rather it is matter of trying to find truths—essential realities. One of the truths that I think we can all agree on is that we don't know how forests work. If we recognize that, then we are left with a choice between a conservation ethic and an exploitative ethic.

The conservation ethic says that if we don't know what we're doing—if we don't understand the long-term effects—then we had best proceed very cautiously, refusing to do anything that might be harmful until we are convinced that we will not harm ecosystems. The exploitative ethic says that unless we have absolute proof that it's harmful, we might as well go ahead and do it. The conservation ethic sees whole systems—the exploitative ethic sees compartments.

Not long ago, I saw a good example of the exploitative ethic in the Vancouver *Sun*. In a 1992 interview, a MacMillan Bloedel logging manager was discussing the need to protect the unique whale habitat at Robson Bight—the only place in whole world, that we know about, where orcas enter shallow water to feed and rub on the pebbles on the beach. "I fully agree that the whales need to be protected," he said, "and I have committed that we will shut down our operations if there is scientific proof that our activities are harming the whales." By the time there is enough proof to convince MacMillan Bloedel, it will probably be too late for the whales. The exploitative ethic sees forests as commodities, largely logs, and is unable to accept that old growth forests maintain the water in the Tsitika River, which maintains the rubbing beaches at Robson Bight.

After pleading with the public to be objective, industrial and government timber managers usually add that what we are faced with, really, is "a new set of values." For example, according to timber industry consultant Patrick Armstrong in a May 1992 Vancouver *Sun* interview, "Forest land use conflicts in this country are predominantly the result of changing social values." In the introduction to the BC Forest Resources Commission 1991 report, Commission Chairman Sandy Peel says, "Driving the current level of dissatisfaction [with forest management] is a dramatic shift in society's values." The Peel report also maintains the forests of BC were "once valued only for their economic worth."

All this is nonsense. People have always valued clean water, clean air, fish, animals, plants, beauty, tranquility and the spiritual gifts of the forest. Suggesting that these values are merely fashionable is an insult, particularly to Indigenous people. The apparent "shift in values" is no more real than "objectivity" or "impartiality." Society's values have not changed, any more than industry's values have changed. There are simply many more people in the game, while the playing field continues to shrink. To suggest cosmetic changes to forest practices in order to appease a social fad is to misrepresent and trivialize an ecological crisis.

Damage to the Forest . . . Including Human Society

A forest is so much more than trees—or timber. A forest is an interconnected web whose fragile strands are composed of rock, soil, water, light, climate and a community of life: animals, microorganisms, fungi, and plants, including shrubs and herbs as well as trees. Figures 1a and

Figure 1a: Fully functioning landscape.

1b show two familiar views of an unmodified forest web as maintained by an unmodified forest. Figure 1a is a "panorama shot" of a fully functioning landscape and 1b is a "close-up" of a fully functioning stand. The trees of a forest depend, for their life and health, on the other components in the forest web; other forms of life in the forest depend, likewise, on the trees. Nothing is disposable in a forest.

If we alter an element in the web, the web itself is altered. If we alter a sufficient number of elements in the web, the web is destroyed. Conventional foresters typically practise timber management by cutting down all the trees on a site, hauling the largest tree bodies away, burning the slash, killing natural shrub/herb growth and planting new trees. The forest is damaged through soil degradation, altered microclimates, fragmentation, pesticide accumulation, and loss of animal, plant, and microorganism habitats and populations. In other words, conventional "forest" practices alter or remove most of the elements in the forest web, but replace only one—the trees. The trees that are planted may grow (in the short term), but the forest is gone. We are

Figure 1b: Fully functioning stand.

left with another ecological equivalent of a prison—a highly selected population, assembled for the purpose of "rehabilitation," but expected to live without the benefit of a normally functioning community or an appropriate environment.

Figures 2a and 2b show two increasingly familiar views of the damaged forest web that follows conventional forest practices—2a is a "panorama shot" of a degraded forest landscape and 2b is a "close-up" of a degraded forest stand.

The timber industry in North America commonly defends practices such as clearcutting by saying they mimic nature. It is true that natural disturbances such as fire, wind and disease epidemics can kill most of the trees on a site, but they rarely kill all of them. Usually, significant numbers of trees, singly and in groups, remain alive.

Furthermore, natural disturbances are random and widely separated in both time and space. They do not occur on planned cycles in perfect rectangles. And lastly—the most obvious point of all—no natural disturbance loads all the dead tree bodies on a truck and hauls them

Figure 2a: Degraded landscape (conventional modification).

off to a mill. Natural disturbances leave the tree bodies on the site, to furnish soil, nutrients and habitat for the next generation of the forest community.

The problem that transcends all others in terms of damage to forests and to human society is the exponential loss of biological diversity—the genetics, the species, the communities, and the landscapes which form the basis for life and for sustaining Earth as we know it. As a result of our human activities, this basis for life is now disappearing at a rate several thousand times the average rate of natural extinction. At the current rate of a loss—the direct result of human exploitation aided by technology—15 percent of the world's species could be gone in ten years. We are literally like the frog, drinking up the pond in which it lives.

We don't even know what we're losing. According to many scientists, we have identified less than ten percent of the species and their associated genetics that keep this planet ticking. We know even less about the functioning of these organisms, their interrelationship with

Figure 2b: Degraded stand (conventional modification).

other organisms, and their singular and collective contributions to maintaining Earth.

We do have some evidence at hand about the results of tampering with ecosystems. The European forestry experience has demonstrated that, in the absence of a fully functioning forest community, trees eventually—after a generation or two or three—weaken and lose vigour. Plantation forests in Europe today are stressed by diseases and insects that once posed no threat to the survival of natural forests.

Society, for its part, is damaged by contamination or loss of water supplies, health hazards associated with pesticides and slashburning, unstable employment patterns, loss of spiritual values, foreclosure on the just settlement of land questions, and the conflicts that inevitably arise when the legitimate needs of all ecologically responsible forest users are unmet.

Overall, the damage to society is another result of the loss of biological diversity. Our culture, including our economy, depends on protecting, maintaining, and restoring healthy, diverse forests. Without biologically diverse, fully functioning forests at all scales, our society becomes like a tree plantation—stressed and vulnerable.

One of the most subtle, yet far reaching kinds of damage to society resulting from industrial timber management is the disempowering of local communities and individuals. Our society is characterized by centralized control of virtually all aspects of life, from what is taught in public schools and universities to how the forests around local communities are used. We are told that clearcuts renew decadent old growth. We are told that plantations are forests. We are told that local decisions must be guided by the dictates of the global economy— an economy based upon destruction of local communities and ecosystems. Disempowering communities is a convenient, effective strategy for corporate interests, and a humiliating, degrading experience for local people.

To initiate genuine changes in our forest practices, we all need to understand that we have been disempowered. In this way, we have all become part of the problem. In addressing the politics of forest use, we need to act, not react. The responsibility for implementing ecologically responsible forest practices and balanced forest use rests with us.

The Basis of New Forest Practices

Change Begins with People

Forests didn't get us into this mess—people did. Thus, collectively society must develop solutions that work for all, recognizing that our future is inextricably linked to the health of the forest. We each have special gifts to offer and no person's gift is any less important than anyone else's gift to making change work.

Our attitude toward power is significant. When I speak to local groups who are concerned about forest use in their community, I tell them, "You have as much power as you believe you have. If you believe you are powerless, you are powerless. If you believe you have power, the power is yours." All of us, and all of our communities, need only open our minds and hearts to the idea that change is urgent, that change is possible, and that change begins with us.

I remember the community meeting at Anahim Lake which launched the West Chilcotin Community Resources Board. One of the primary reasons for establishing this Board—comprised of Indigenous people, ranchers, wilderness tourism operators, trappers, small loggers, and community members—was to shift power for planning use of the local forest from government and the timber industry to the local community. Among other issues, local people wanted an immediate reduction in the AAC in the area and immediate incorporation of local concerns into logging practices. The public meeting at which the Board was formed was observed by several Ministry of Forests staff members and industrial foresters who had driven four hours from Williams Lake in order to be present.

After the resolution establishing the Board was passed by the community, a Ministry representative stood up to congratulate the group and to encourage them to send a representative to the next Ministry of Forests planning process meeting. His remark was followed by a long silence in the hall. Then a tall rancher unfolded himself from his chair and replied: "I guess you don't understand what happened here today. We aren't going to come to your meetings any more, but perhaps you can come to our meetings . . . when we invite you."

In evaluating any future directions in forest use and forest practices, probably no other indicator will predict the success as reliably as the extent of ecologically responsible community control over decisions, practices and standards. Edward Grumbine, in his excellent new book *Ghost Bear*, reminds us that effective community control is a key aspect of stewardship:

Anthropological research worldwide has shown that there existed many varieties of commons management prior to the rise of commodity-oriented societies. The two outstanding features of most successful, long-term cultures that worked with nature were intimate knowledge of plants, animals, and ecosystems and small-scale community control.

Over the last hundred years, centralized government, centralized education and single interest control of the forests have removed responsibility for the local forest from the local communities that depend on them. Decisions that shape the destiny of small communities all over BC are frequently made by government bureaucrats in Victoria or by corporate executives in Vancouver, Toronto or Tokyo. In these cases, responsibility is usually directed toward the next election or the next quarterly profit and loss statement. The forest is seen only on coloured maps that "integrate" other forest uses with logging, or on computer printouts of log and lumber values. People are known only as employment statistics. Problems come in envelopes and leave in envelopes. Responsibility to communities and ecosystems gets lost in authorless annual reports and faceless inter-office memos. If we are to achieve the two-fold objective of ecologically responsible forest practices and balanced forest use, I believe we must start with responsible community control of the forest—understanding that what we do to the forest we do to ourselves.

Such control supplies the sensitivity, the responsiveness and the long-term view of people—not bureaucracies or corporations, but people who have a commitment to being a part of the forest and a stake in using the forest in ways that maintain its full integrity.

Thinking Like a Forest—A New Definition of Forest Practices

The first job before us is to define forest practices in a simple, honest way—namely, as any human activities that affect the forest. If we understand that a forest is more than logs standing vertically, then we can no longer equate forest practices to conventional timber production practices. The phrase "changing forest practices" will no longer mean finding better ways to cut trees or kill brush or design plantations. We can move to ecosystem-centered practices instead of human-centered approaches.

Exploitation of forests for timber and "enhancement" of forests to produce more timber characterize the old models. This is a forest *insensitive* approach that focuses on what to take for the immediate benefit of *some* people. In contrast, protection, maintenance and restoration (where required) of forests are the cornerstones of an ecosystem-centered model. This is a forest sensitive approach that focuses on what to leave—fully functioning forests at all scales, for the sustenance of all people.

An ecosystem-centered approach is based on the clear understanding that what we do to the forest we eventually do to ourselves. Therefore, the first priority of ecosystem-centered forest use is to maintain the ecosystem in a fully functioning, healthy condition. (Look again at Figures 1a and 1b.) The means for achieving this are not difficult to understand, and I will discuss these later.

Basic to the goal of maintaining fully functioning forests are three simple concepts: the time/space challenge, keeping all the parts, and water as the connector.

The Time/Space Challenge

Human beings and forests operate on vastly different scales of time and space. Forests function on continuous cycles of 200, 500, 1500 years or more. Our current attempt to shorten this continuum so that we are cutting and planting trees on 60-, 80-, or 120-year cycles or "rotations" short-circuits the natural pattern of the forest ecosystem. The time problem is further complicated by corporations, governments and other human institutions that interface with the forest, but operate on yearly budgets and monthly profit and loss statements. Human beings in our culture find it very difficult to relate to forest time. It is interesting to note that "normal" industrial timber rotations are roughly equivalent to the time span of a long or very long human life, even though trees have a natural lifespan that is two or five or ten times that long.

We encounter the same problem with space. A forest ecosystem is not characterized by sharp boundaries, straight lines, right angles and tidy compartments. Forest ecosystems exist on many spatial scales simultaneously, from the habitat of microscopic organisms to major watersheds, continental drainage systems and global climate patterns. The patterns and processes found at each scale are interdependent and interconnected with all other scales. Yet when timber managers plan forest use, they operate in human terms—machine limits, hauling distances, and arbitrary cut-block boundaries.

If we are designing forest practices that maintain fully functioning forest ecosystems, then we will face major challenges in terms of time and space. In fact, if we do not feel challenged in that way, we are probably overlooking something critical. In order to practise ecosystem-centered forest use, we must learn to "think like a forest."

Keeping All the Parts

Another concept fundamental to maintaining fully functioning ecosystems is very intuitive: keep all the parts. When we use the forest for any purpose, let's make sure that we don't destroy or discard anything that is necessary to maintain fully functioning forests.

This idea of keeping all the parts is one that most of us understand only too well when we think about technology and its products. If we take apart our watch or our camera or our car, we know that we have to put all the pieces back together in the same way they came out, or the watch or camera or car is not going to work the way it's supposed to. It doesn't matter whether or not we understand what function any particular piece serves. We simply know better than to say, "I don't know what this gizmo does. It doesn't look so important. I'll just throw it away."

We already know that with any mechanical device, we have to protect all the parts, maintain them in good working order and, as a last resort, restore damaged parts so that they function effectively. I think we can learn to transfer these understandings to ecosystems—to realize that we are first and foremost citizens of the natural world, that we depend for our very lives on the ecosystems of the earth, and that the same principles apply.

This means we need to have the humility to admit that we don't know how forests work. In the past, we have concentrated our research and our economic activity on the most obvious part of the forest web, ignoring the complicated mutual relationships that trees have with thousands of other organisms, from bears to bacteria. For example, microscopic organisms are present in forests in staggering numbers. There are millions of these organisms in every single gram of forest soil. Most of them have never been studied or even identified. We don't yet understand what all those fungi and bacteria are doing there. We may never understand. But that doesn't mean it's safe to say, "I don't know what this organism does. It doesn't look so important. Let's just get rid of it."

Yet, every time we clearcut a forest, we are changing the habitat, the surrounding temperature, the community, the soil structure and the moisture regime for these microorganisms. We have no clear idea how many survive, for how long, or even the condition of the survivors.

One kind of microorganism that we know a little about are the mycorrhizal fungi that live in a symbiotic relationship with trees and other plants on their roots. In exchange for drawing sugars and carbohydrates from a tree's system, mycorrhizal fungi extract water and nutrients from the soil and make it available to a host tree. They also form a physical shield that protects the roots from root decaying fungi.

Associated with some mycorrhizal fungi are bacteria that extract nitrogen from the atmosphere and make it available to an associated tree or other plant. Nitrogen is one of the fundamental building blocks of life, a component of protein. No plant can survive without nitrogen, yet the nitrogen in the atmosphere is not available in a form that trees can use. As a result, bacteria associated with mycorrhizal fungi furnish an important source of nitrogen for trees in BC and elsewhere.

There are probably thousands of species of mycorrhizal fungi, each adapted to a particular tree species, a particular stage in a tree's life, or a particular local condition. We know that mycorrhizal fungi are essential to healthy tree growth and that clearcutting, slashburning and pesticides destroy them or drastically impair their function. I don't think we need to wait for a lot of scientific research to tell us more about mycorrhizal fungi before we start cutting trees in a way that doesn't affect them profoundly.

In this example of keeping all the parts, I want to be clear that I'm not saying "Do nothing." We don't have to stop cutting trees and stop using the forest for fear that we'll step on a fungus. Ecosystems have a remarkable characteristic that human-created objects lack: redundancy. Ecosystems are designed so that the same need can often be met in a variety of ways, thus providing several fail-safe mechanisms in the event of a major disturbance such as a fire, insect infestation or wind damage.

This quality of redundancy means that *sometimes* we can remove *some* parts and still maintain a fully functioning ecosystem. What we can't do—if we want to maintain the ecosystem—is destroy the redundancy or reduce it to levels that impair the short-term and/or long-term functioning of the system. This is what current forest practices do in many ways, including the elimination of old growth from the forest landscape, the removal of all old trees and snags by clearcutting, and the elimination or truncation of the shrub/herb stage of forest growth through pesticide applications and other methods of "brush control." When we accept redundancies in the forest as our basis for using parts of the forest, we need to do so with humility.

If we applied the same logic to a group of human beings, we would declare certain individuals redundant to maintaining a fully functioning human population. Not too many of us would volunteer to be redundant.

Water is the Connector
The third concept that can guide our efforts to maintain fully functioning forest ecosystems is the idea that water is the connector in the forest. Water is everywhere in a forest, from every living cell, every rain droplet and every little rivulet up to major river systems and wetlands. Three aspects of water—quantity, quality and timing of flow—provide us with clear ways to evaluate how careful we are in our forest use. If we maintain these aspects in an ecosystem, from the smallest rivulet or stream up to the largest river system, we are very likely doing a good job of maintaining the rest of the ecosystem's functions. However, in order to do this, we must recognize that while water influences all parts of an ecosystem and is an absolute requirement for the living parts, other parts of the ecosystem also affect water. For example, without decaying wood from large old trees in the forest, soil water filtration and storage is degraded. The interdependence of parts must always be respected if our activities are to maintain fully functioning forests.

The Scientific Basis of Wholistic Forest Use
We must beware of advice recommending that, before we implement new ideas, we need to do more studies, convene another Royal Commission, run another survey, launch a major research effort, or another panel or committee or open house. We've had enough research, enough discussion, enough coffee and doughnuts.

I'm not against research and discussion, but it seems clear to me—on this issue—that the time to study the question is long past. We know that we need to change, we know enough to begin the change, and I think we're smart enough to learn as we go along.

Remember that no one ever bothered to research the ecological or long-term economic virtues of clearcutting. No one ever analyzed the costs and benefits of exporting pulp and two-by-fours rather than photographic paper and violins. We just did it. I'm not suggesting that we stop doing research, but I am suggesting that we know enough to change. Now.

We understand how to protect biological diversity, how to protect water and soil, how to carry out selection systems of timber cutting, and how to use soft technology and skillful marketing. We understand solving our problems with ecologically responsible finesse, not simply using more machine power. We can let our experience as we change guide future research, and that can help us refine our changes as we go. One tool that can help us achieve ecosystem-centered and ecologically responsible forest practices is the relatively new science of landscape ecology.

Landscape Ecology
Stands of trees do not function in isolation, any more than individual trees do. At the large or landscape scales, the forest is composed of clusters of interdependent, interconnected ecosystems or landscape patches, from stands of young trees and stands of old growth to creeks, wetlands, and meadows. Energy is exchanged among these clusters through climate, soil and water. Protecting the natural pattern or mosaic of any landscape is necessary, both to maintain the overall viability of the landscape and to maintain the health of each individual ecosystem.

The study of landscape ecology is concerned with the ecological functioning of entire landscapes over both space and time. The linkages among interdependent ecosystems or stand units are critical to the study of landscape ecology: actions and events at any level will have effects throughout the landscape, on both smaller and larger scales. As landscape ecologists Richard T.T. Forman and Michel Godron put it, "An action here and now has effects there and then."

Forest landscapes are like very large waterbeds. If you push down here, the ripples will pop up again somewhere else. The question is, where, when, and how? Connections across the landscape must be respected to ensure the long-term viability of forests that make up the landscape. Landscape ecology seeks to understand and protect the interconnection of the whole landscape during human use.

Landscape ecology developed as a distinct discipline largely in Europe during this century, originating as an attempt to integrate the spatial concerns of geography with the time-scale concerns of ecology. In Europe, where dense and long-established populations have made the effects of human disturbance on ecological systems obvious, landscape ecology is now viewed as the appropriate scientific basis for land use planning, conservation, resource use and land reclamation. European land managers are aware of the need to examine the effects of their land use plans on entire ecosystems and on the energy and nutrient flows through ecosystems. The discipline has now expanded to include aspects of sociology, psychology, economics and cultural studies.

The science of landscape ecology thus operates on the vital larger scale, recognizing the landscape as the framework within which standlevel ecosystems function. Landscape ecology is the logical starting place for all forest planning and use. However, this is a much larger view than considered necessary by many scientists, foresters and planners. In fact, this practical science has been largely overlooked in forest planning in Canada because timber managers have, for the most part, a very limited understanding of landscapes, landscape ecology, and the need for landscape planning and management.

The *British Columbia Ministry of Forests Policy Manual* defines "Forest Landscape Management" as:

> The activity by which *visual* and *aesthetic* landscape values are identified, inventoried and analyzed, and are protected or enhanced, according to their relative importance, within integrated resource and management plans, and during resource development. [Author's italics]

Conventional Canadian timber managers perceive "landscape" to mean "pretty scenery." They use the concept of "landscape planning" to refer to visual resources. Recently, draft biodiversity guidelines, which incorporate principles of landscape ecology, have been released by the BC Ministry of Forests. However, these guidelines have been largely ignored by the timber industry and even the Minister of Forests has directed that we "go slow" in implementation. If the Minister also wanted to "go slow" with logging while dragging his feet with the biodiversity guidelines, then implementing landscape ecology in forest practices would be possible. However, timber cutting proceeds apace, largely in sensitive forests where sustainable timber management is questionable, regardless of the way we do it. Virtually all of the highly productive, durable sites where sustainable timber management makes sense have been logged. Application of landscape ecology lags well behind this final green gold rush, and we are likely to end up with only landscapes to restore, not landscapes to protect and maintain. In other words, things are not getting better. They are getting worse faster.

Under conventional clearcut logging methods and planning priorities, the larger expanses of the forest ecosystem—forest landscapes—are often fragmented, critical landscape connections are broken, habitats are disrupted or destroyed, energy and water flows are interrupted, movements of animals (and plants whose seeds are carried by animals and water) are obstructed, and non-timber forest values such as wilderness, water and balanced allocation of human uses are put at risk. Overall, the most significant impact of fragmentation is the loss of biological diversity required to sustain forests and all ecosystems at all scales.

I would propose that the validity of forest practices in the future must be measured against the principles of landscape ecology. These principles provide us with a strong and significant measure of the integrity of any forest practice. Landscape ecology is the science that describes and advocates *protection for forest connections in time and space.* The field of landscape ecology is not easy to summarize in this short space, but a brief explanation of four principles of landscape ecology—time, space, connectivity, and diversity—will help, I hope, to demonstrate that our future forest practices must respect these principles on all levels and at all times.

• Time and Space
Earlier I mentioned the challenge that humans face attempting to comprehend "forest time" and "forest space." Because it incorporates a basic outlook that is both "long term" and "far reaching," landscape ecology provides an effective means of addressing this challenge. Landscape ecologists recognize that the total effect of any disturbance or human activity extends well beyond the season, year, or decade of the event, and that a disturbance on any given site will be reflected in the dynamics of the entire landscape for the life of the ecosystem. The effects of any given disturbance may be ameliorated by time or dis-

tance, but they are not erased. For example, a regenerated clearcut may begin to stabilize soil, but it does not begin to provide the functions of the old growth forest that once occupied this part of the landscape. Because of likely local species extinction (particularly microorganisms) that result from clearcutting, development of an old growth forest in this part of the landscape again is unlikely—no matter how long we wait.

In terms of *space*, the forest landscape functions on many levels—from millimetres of soil, patches of rocks, and stands of trees, to entire watersheds and beyond. In terms of time, each phase of the forest plays a crucial role in maintaining a stable, diverse forest landscape. Shrub/herb and old growth phases are critical for capturing and storing nutrients for future forest growth. The shrub/herb phase often has the largest species diversity of any phase, but old-growth has specialists not found in any other forest phase. Trees in young and mature forests produce wood fibre at the most rapid rates, while old growth forests provide necessary habitat for animals, including the insects and birds that prey on forest "pests." Old growth, the most stable and often the most abundant forest in a healthy forest landscape, is also the earth's most important land-based storage system for carbon—a function critical in regulating atmospheric levels of carbon dioxide ("greenhouse gases").

• Connectivity

Alongside time and space, a third important concept of landscape ecology is *connectivity* within the landscape. Diverse "patches" or habitats for various plants, animals and microorganisms are required to maintain an ecosystem. However, these patches are valuable only if they are connected to one another in some way. Connectivity within a forest landscape is provided by movement corridors, which are frequently riparian ecosystems. A riparian ecosystem includes the *riparian zone* or wet forest adjacent to a stream, river, lake, or wetland, and the *riparian zone of influence* or upland forest immediately upslope from the riparian zone.

Riparian ecosystems are of special importance as landscape connectors. They are arranged in a branching network that extends throughout a forest landscape, and they contain varied but repeating patterns of plant and animal habitat. Because of their wet and diverse nature, riparian ecosystems (particularly riparian zones) frequently survive large natural disasters such as fire and windstorms. As movement corridors, riparian ecosystems provide migration routes for large and small animals. For example, large ungulates such as moose and elk use these corridors to migrate between seasonal ranges. Simply maintaining suitable habitat patches is not sufficient to guarantee persistence of some species. Maintaining migration corridors is necessary to prevent large scale extinction. Mammals and birds move many plant seeds around the forest landscape, and some seeds are also carried by water. Thus, plant dispersal routes also tend to follow riparian corridors.

Riparian ecosystems are connected from valley to valley by treed slopes that function as forest corridors running up and down forest slopes. These cross-valley corridors provide routes for many animals and plants to move back and forth between riparian ecosystems and upslope habitat patches, ranging from mid-slope old growth forests to alpine areas.

Groundwater is another type of landscape connector that transports and disperses nutrients and energy both within forest patches and throughout the forest landscape. These water flows are concentrated and cycled in riparian ecosystems, where they nourish the most lush and diverse species populations found in most forest landscapes. Eventually, nutrients and energy are released from the riparian community into the aquatic ecosystem, which then carries nutrients and energy to distant parts of the landscape.

Riparian ecosystems, cross-valley corridors and groundwater are major examples of *connectivity in space*. Blocked energy/nutrient flows can lead to ecosystem impoverishment. A large-scale example of this is the effect of the W.A.C. Bennett hydroelectric dam on the Peace River. The dam prevents nutrients and silt from reaching the formerly productive Peace-Athabasca delta, resulting in the loss of specific habitat types that previously helped support bison, and likely other species as well. This same effect occurs with every road and clearcut, and is cumulative across landscapes modified by logging.

Natural forest stages are examples of *connectivity in time*. From the shrub/herb phase through the young and mature forest phases to old growth, each stage has an important role in maintaining a healthy and diverse forest landscape. If the connectivity in time is broken by, for example, eliminating the shrub/herb and old growth phases, the result is fragmentation and disruption of many other interlocking patterns within the ecosystem. It is like removing links from a chain.

In a forest ecosystem, impacts on any one part of the landscape cannot be isolated, but will affect all parts in some degree at some time. Impacts which reduce or break natural landscape connectors will have direct impacts on animal, plant, energy, nutrient and water movements. Also, because of landscape connectivity, events or conditions in one part of a landscape will affect the ecology of areas well beyond the physical boundary and time of those events or conditions. Water pollution and air pollution are the most familiar examples of this principle, and other examples are easy to find. Erosion from landslides caused by clearcutting results in far-reaching downstream impacts on fish populations. Smoke from slashburning impacts large landscapes, affects human health (and likely the health of other species), and contributes to global warming. The effects of most pollution extend well beyond the point source of the pollutant, both physically and ecologically.

• Diversity
A fourth basic concept in landscape ecology is *diversity* or *heterogeneity* within a landscape. A natural forest landscape, for example, includes a variety of habitat patches containing diverse mixtures of trees, shrubs, herbs, animals and microorganisms. Habitat patches are shaped by unpredictable natural disturbances, from the falling of a limb or tree to a wildfire, and vary according to moisture, slope, elevation, aspect, soil and other physical and biological factors. This kind of natural diversity is vital to ensure that all the parts are available for forests to function. In contrast to the homogeneity and resulting instability of clearcuts and tree plantations, the heterogeneity of natural forests results in resilient, stable ecosystems.

Landscape heterogeneity, or diversity, is essential to forest landscapes for many reasons. For example, landscape diversity is required for the persistence of animal species that require more than one ecosystem or patch type in order to survive and reproduce. Grizzly bears, for example, require several different ecosystem types for different needs—for summer and fall feeding, for denning and safe rearing of young, and for migration between seasonal ranges. Even setting aside one landscape with a full range of habitat types for grizzlies, as has been done in the Khutzeymateen Valley, may eventually be futile unless *all* habitat types and adequate connections between them are maintained in adjacent landscapes.

Diversity in ecosystems also contributes to redundancy: the ability, mentioned earlier, of an ecosystem to perform important functions in more than one way or to a capacity beyond current needs. Redundancy, maintained by landscape diversity, is a vital fail-safe function that allows ecosystems to survive stress. It also means that species of various kinds can persist, not just exist on the brink of loss—something that provides options for adapting to changing conditions.

If habitat areas become too small to be effective, extreme diversity can result in negative effects. As with so many aspects of forest functioning, a balance is required between diversity and repeating patterns, between heterogeneity and homogeneity. On the whole, however, heterogeneity is a positive feature in most landscapes. Diverse habitat patches provide diverse resources and tend to stabilize landscape processes.

Stand Level Ecology
The stand level, or the individual ecosystem, has been, until recently, the most familiar scale for application of ecological principles. However, the implications of landscape ecology and the need to maintain fully functioning forests at all scales require that we approach stand level ecology in a new way. We can no longer think of stand ecology simply in terms of trees or treed areas that are similar enough to be managed for timber using the same prescription. Stand ecology also covers animals, plants and microorganisms at all scales.

Various communities or ecosystems such as wetlands, prairies and alpine areas are habitat patches or "stands" within a forest landscape. When we define a stand or habitat patch or ecosystem, we cannot be limited by human-scale and human-centered perceptions. The size of a habitat patch suitable for a bacteria is very different from the size of a habitat patch suitable for a squirrel or a fern, which is in turn very different from the size of the habitat patch suitable for a woodpecker or a beaver, devil's club, moose, cougar, Douglas fir or grizzly bear.

As Michael L. Morrison, professor of wildlife biology at the University of California, Berkeley, put it, "Our problem is that we cannot see through the eyes of an animal; we cannot directly know what the animal perceives or what influences these perceptions have on its behavior." If we manage ecosystems according to our own scale perceptions, the results may be suitable for us (for a while), but they will not be suitable for other species.

The principles of landscape ecology imply that stands are basically landscapes at small scales. If we are to use ecosystems in ecologically responsible and balanced ways, then *land use planning must proceed on all scales simultaneously*. The narrow view we exercise as humans must not blind us to the structures, functions and composition of ecosystems at other scales.

Jerry Franklin, an old growth forest ecologist at the University of Washington, sees landscape ecology planning as a sound, practical system for making forest use decisions. But, as he says, it's not much use to him or his colleagues in Washington and Oregon:

> We do not have any full landscapes left where we can apply this. We are looking largely at restoration. We are trying to put landscapes back together again. In Canada, I urge you to use this ecological knowledge and apply it on the ground. If you find out later that you protected too much, if you find out later you have been too conservative in what you have done, it will not be too late to cut the trees down, but it is always too late to stand them back up.

Planning on all scales simultaneously may sound like a formidable—or impossible—challenge. However, as demonstrated below, both stand and landscape ecology can contribute valuable information to forest use planning so that essential "nodes" and "corridors" can be identified and maintained in their natural state, in order to protect connections and diversity through time and space while allowing us to use the forest in an ecologically responsible and balanced way.

New Ways in Practice

Any model of ecosystem-based land use planning must respect two important priorities:

The first priority is *ecological responsibility* in any type of human use. Ecologically responsible plans and activities are those that maintain fully functioning ecosystems at every scale, from the smallest stand or patch to the largest landscape in both the short and long terms.

The second priority is *balanced human and non-human use*. Within the context of ecological responsibility, we must ensure that all organisms, both human and non-human, have a fair and protected land base in order to meet their needs and carry out their functions in the ecosystem. To implement these two priorities in the context of landscape ecology, we can use a sequence of ecosystem based planning, as shown in the flow chart in Figure 3.

Many of the items in this sequence are self-explanatory. However, I would like to explain briefly two of the steps—identifying a network of protected ecosystems and allocating specific human uses.

Identifying a Network of Protected Ecosystems

Maintaining the functioning of the forest landscape means establishing two kinds of protected areas: 1) *large protected reserves*, and 2) *protected landscape networks*. Both of these protected areas accomplish the same primary goals: to protect the full range of biological diversity at all scales and to maintain connectivity in the landscape.

Large Protected Reserves

Large protected reserves give us a way to protect the temporal aspect of landscape ecology. These reserves must be large enough to withstand large natural disturbances while retaining their resiliency and integrity as a fully functioning forest landscape. How large is large enough? Scientists suggest that protected landscapes must be 50 to 200 times the size of the largest anticipated natural disturbance.

Large protected reserves constitute entire drainage basins or watersheds ranging in size from about 5,000 hectares upward. In landscapes shaped by large disturbances, such as fire in the boreal forest, large protected reserves will need to occupy areas of thousands of square kilometres. These areas are the storehouses of biological diversity necessary to maintain all of the forest, both that modified by human use and that which remains unmodified. The more severe and extensive the modifications of the landscape, the greater the need for large protected reserves. Figure 4 is a conceptual diagram of a landscape that includes large protected reserves (labeled as "protected drainage basins").

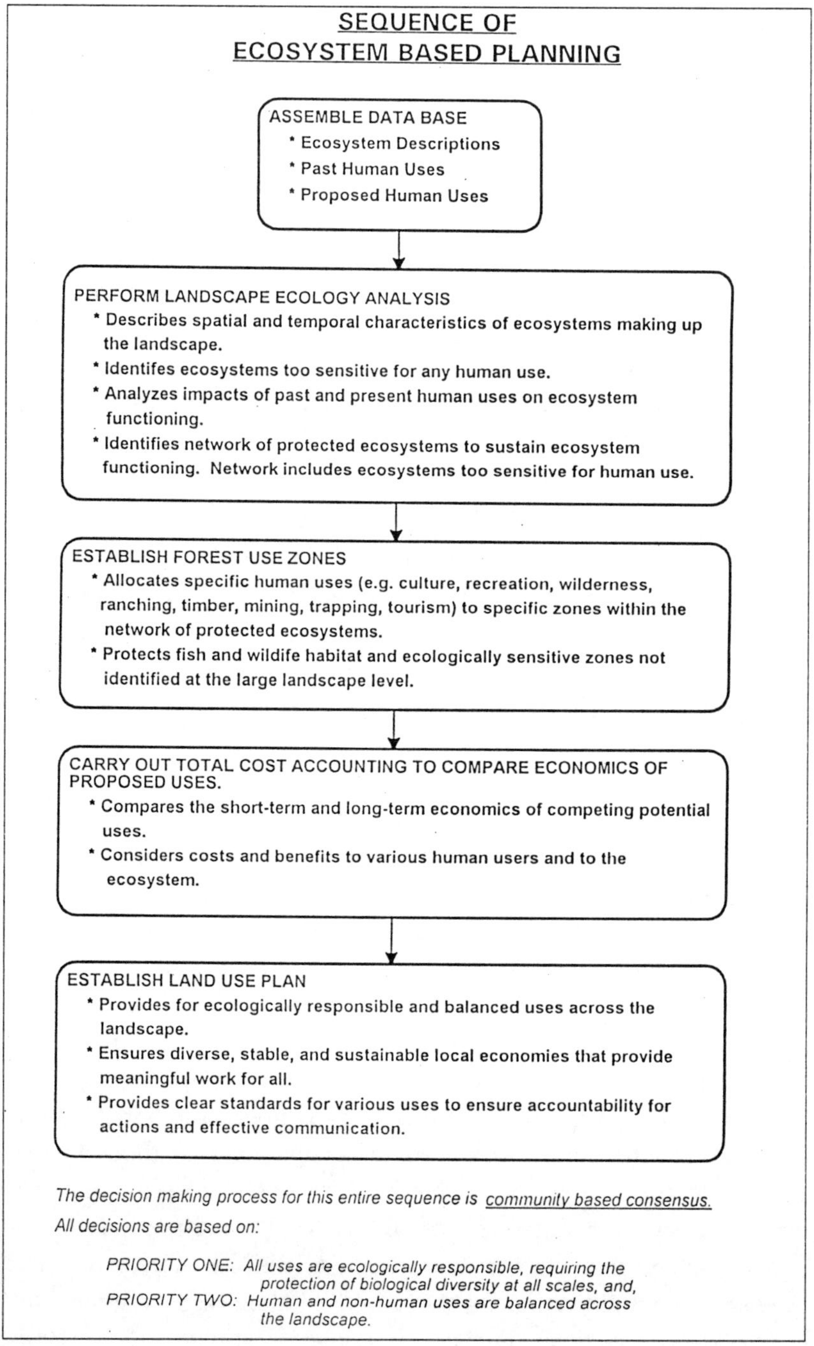

SEQUENCE OF
ECOSYSTEM BASED PLANNING

ASSEMBLE DATA BASE
* Ecosystem Descriptions
* Past Human Uses
* Proposed Human Uses

PERFORM LANDSCAPE ECOLOGY ANALYSIS
* Describes spatial and temporal characteristics of ecosystems making up the landscape.
* Identifes ecosystems too sensitive for any human use.
* Analyzes impacts of past and present human uses on ecosystem functioning.
* Identifies network of protected ecosystems to sustain ecosystem functioning. Network includes ecosystems too sensitive for human use.

ESTABLISH FOREST USE ZONES
* Allocates specific human uses (e.g. culture, recreation, wilderness, ranching, timber, mining, trapping, tourism) to specific zones within the network of protected ecosystems.
* Protects fish and wildife habitat and ecologically sensitive zones not identified at the large landscape level.

CARRY OUT TOTAL COST ACCOUNTING TO COMPARE ECONOMICS OF PROPOSED USES.
* Compares the short-term and long-term economics of competing potential uses.
* Considers costs and benefits to various human users and to the ecosystem.

ESTABLISH LAND USE PLAN
* Provides for ecologically responsible and balanced uses across the landscape.
* Ensures diverse, stable, and sustainable local economies that provide meaningful work for all.
* Provides clear standards for various uses to ensure accountability for actions and effective communication.

The decision making process for this entire sequence is <u>community based consensus.</u>
All decisions are based on:

PRIORITY ONE: All uses are ecologically responsible, requiring the protection of biological diversity at all scales, and,
PRIORITY TWO: Human and non-human uses are balanced across the landscape.

Figure 3.

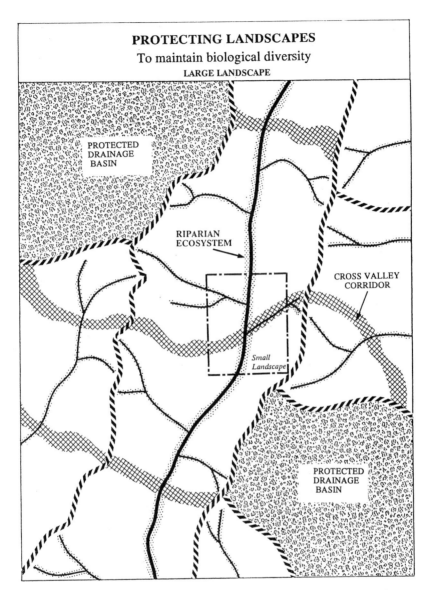

Figure 4.

In order to maintain healthy landscapes, large protected reserves cannot exist as islands, regardless of their size. They must be connected across the landscape as shown in Figure 4. Ideal connectors include *riparian ecosystems* and *cross-valley corridors*.

Riparian ecosystems include the riparian zone (wet forest area along creeks, rivers, lakes, wetlands, and all water bodies) and the riparian zone of influence (the upland forest immediately adjacent to the riparian zone). Cross-valley corridors consist of wide bands (300 to 5,000 meters) of forest that provide valley to valley connections between large protected reserves and across the whole landscape. Cross-valley corridors should be made up of representative ecosystems found in the landscape, and should not be blocked by land features such as cliffs and rock bluffs that obstruct the movement of large mammals.

Protected Landscape Networks
In the areas between large protected reserves, ecologically responsible human modification (wholistic forest use) may occur. However, it is still necessary to further protect the landscape's framework in the areas outside of large protected reserves. This is achieved through protected landscape networks. As shown in Figure 5, riparian ecosystems and cross valley corridors are also components in this level of landscape protection, along with other types of smaller protected areas.

The term network makes clear the idea of connectivity; however, we need to keep in mind that the corridors and protected clusters are not narrow strips, but broad areas encompassing complete ecosystem types. The overall area, seen from above, resembles a piece of Swiss cheese with wide connecting corridors of protected ecosystems interrupted by "holes." The protected landscape network forms the framework within which (in some of the holes of the Swiss cheese) various ecologically responsible human activities may occur. The complete set of components needed to maintain the landscape framework, and therefore part of a protected landscape network, includes:
— Riparian ecosystems—the backbone of the network
— Representative ecosystems—small protected nodes of 400 hectares or more. Smaller protected nodes are established where an ecosystem type is strategically located, rare or endangered. Old-growth or late successional forests are particularly important representative ecosystems to include in the network.
— Sensitive ecosystems—shallow soils, very wet and very dry areas, steep and/or broken terrain
— Cross-valley corridors

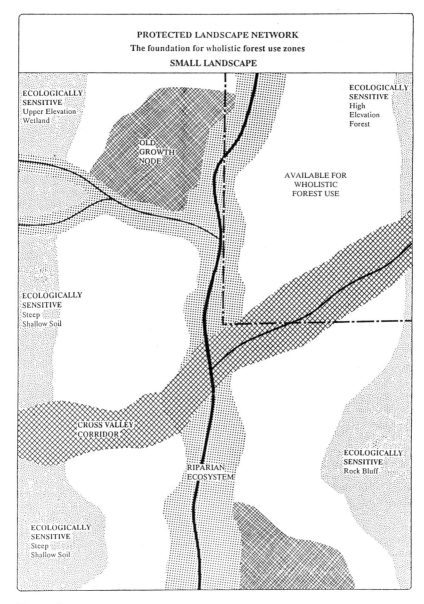

PROTECTED LANDSCAPE NETWORK
The foundation for wholistic forest use zones
SMALL LANDSCAPE

ECOLOGICALLY
SENSITIVE
Upper Elevation
Wetland

ECOLOGICALLY
SENSITIVE
High
Elevation
Forest

OLD
GROWTH
NODE

AVAILABLE FOR
WHOLISTIC
FOREST USE

ECOLOGICALLY
SENSITIVE
Steep
Shallow Soil

CROSS VALLEY
CORRIDOR

ECOLOGICALLY
SENSITIVE
Rock Bluff

RIPARIAN
ECOSYSTEM

ECOLOGICALLY
SENSITIVE
Steep
Shallow Soil

Figure 5.

Allocating Human Uses: Wholistic Forest Use Zoning

We know that establishing large protected reserves and protected
landscape networks will protect the landscape ecology—the patterns
and connections—of the forest. Within the framework provided by

the protected landscape network, ecologically responsible and balanced human use is achieved by establishing wholistic forest use zones.

In establishing wholistic forest use zones, the most sensitive and easily damaged human uses are accommodated before allocating areas for more aggressive human uses.

Criteria for various zones are based upon ecological, social and economic factors. Priority is always given to the protection of ecological/natural factors, because societies and economies are based on ecosystems, not the other way around.

In order of their establishment, typical forest use zones include:
— Cultural/Spiritual—areas which are culturally or historically important to local people
— Ecologically Sensitive—small sensitive areas not identified or protected in the protected landscape network
— Fish and Wildlife Habitat—small necessary habitat not identified or protected in the protected landscape network.
— Recreation-Tourism-Wilderness
— Wholistic Timber Management

Within their respective zones, human uses must be carried out in accordance with ecologically responsible standards. While a particular use may have priority within a given zone, this does not preclude other uses from occurring, provided these uses do not damage or otherwise prejudice the priority use. Some uses, such as adventure tourism and timber extraction, are mutually exclusive and would not occur in the same zone. While timber extraction is a particularly aggressive use of the forest, when carried out in an ecologically responsible manner, it can be combined with many other uses. Once human uses are allocated, a wholistic forest use zone map might look something like Figure 6.

Zoning for Timber: Wholistic Timber Management Zones
Ecologically responsible timber management is essential since timber extraction will likely continue to be a major use in many forest landscapes. However, timber extraction under wholistic forest use looks different in almost every respect from conventional timber management. Compare the landscape and stand views in Figures 7a and 7b with the same views in Figure 2.

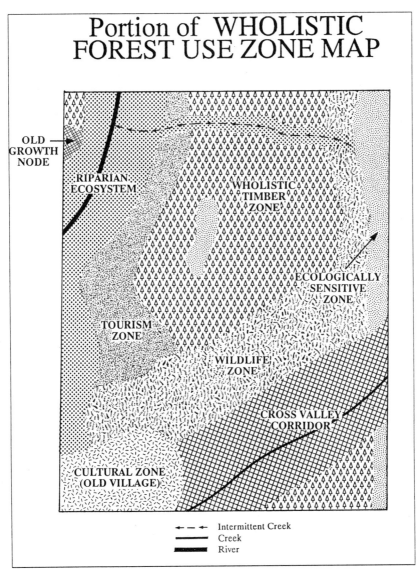

Portion of WHOLISTIC
FOREST USE ZONE MAP

OLD GROWTH NODE

RIPARIAN ECOSYSTEM

WHOLISTIC TIMBER ZONE

ECOLOGICALLY SENSITIVE ZONE

TOURISM ZONE

WILDLIFE ZONE

CROSS VALLEY CORRIDOR

CULTURAL ZONE (OLD VILLAGE)

Intermittent Creek
Creek
River

Figure 6.

Specifically, ecologically responsible timber management in wholistic timber management zones will:

1. Maintain composition and structures to support fully functioning forests. Important forest structures such as large old trees, snags, and large fallen trees are maintained by letting a minimum of 20

Figure 7a: Landscape with wholistic modification.

    to 30 percent of overstory trees (well distributed spatially and by species) grow old and die in any timber extraction area.

2. Use ecological rotation periods of 150 to 250+ years.
3. Prohibit clearcutting as currently practised and utilize ecologically appropriate partial cutting methods that maintain the canopy structure, age distribution and species mixtures found in healthy natural forests in a particular ecosystem type.
4. Prohibit slashburning.
5. Maintain/restore fire necessary for ecosystem functioning.
6. Allow the forest to regenerate trees through seeds from trees in the logged area. Tree planting will generally not be required because a diverse, fully functioning forest is always maintained.
7. Maintain ecological succession to protect biological diversity. The process of brush control will be avoided. Over time, all forest phases must occupy every forest site, even on sites managed solely for timber.
8. Prohibit pesticide use. Disease, insects and shrub/herb vegetation

Figure 7b: Stand with wholistic modification.

are essential parts of a fully functioning forest.

9. Minimize soil degradation by:
    a. first, avoiding use of roads (includes skid roads) wherever possible, and
    b. constructing narrow roads (includes skid roads) that are fit into the terrain.
10. Protect water by:
    a. protecting riparian ecosystems at all scales,
    b. minimizing alterations to natural drainage patterns, and
    c. deactivating old roads (includes skid roads) to reestablish natural drainage patterns.

Wholistic timber management—also called ecologically responsible forestry or "new forestry"—means practicing timber management as if forests mattered. It means that forests would always occupy sites where timber is being cut. It means, while paying close attention to standards at the stand level, also managing timber (and all other human uses) within the context of protecting landscape ecology.

Protection, Maintenance, and Restoration

If we are to have fully functioning forest ecosystems today and in the future, our ideas about forest practices must expand to include three broad categories of human activities: forest protection, forest maintenance and forest restoration.

By forest protection and forest maintenance, I simply mean keeping all the parts—from the smallest stand to the largest landscape. By forest restoration, I mean doing our best, with the limited knowledge we have, to help nature reestablish good ecological functioning in the forest areas that we have degraded in the past. Restoration is the price of exploitation and will be a major obligation facing forest users in the future.

I would propose that forest restoration must be adopted as a key forest practice immediately. Important measures of the adequacy of any forest use plan would be the extent of forest restoration included, and the standards offered to ensure ecologically responsible forest practices so that forest restoration will not remain forever on the list of essential forest practices.

Appropriate forest restoration requires that we shift our approach from redesigning natural systems to following nature's lead. Two ways of doing this are allowing natural fires and taking a smaller scale approach to forestry.

As long as adequate sized large protected reserves exist, and adequate old growth forests are protected from destructive human uses within the managed landscape, natural fires can usually be allowed to run their course in large protected areas. In this way, biological diversity at the landscape level would be maintained according to nature's plan. In drier forest areas, of course, this principle would have to be applied judiciously. After 50 years of sophisticated fire suppression, many of the drier forests in BC have accumulated unnaturally large amounts of fuel. It would be necessary, in some locations, to set fires in small areas at the wettest times of the year, or otherwise reduce fuel loads (by thinning, for example) to more natural levels and to break up fuel continuity in the landscape.

The scale of any forest practice is a good way to gauge ecological responsibility. A clearing of one or two hectares that retains large old trees is completely different from a 50- or 100-hectare clearcut. Planting some trees among predominantly natural regeneration in order to help restore lost diversity is completely different from planting millions of trees in order to convert old growth forests to plantation monocultures. Carefully removing a few selected old growth trees for local use in the manufacturing of wood products such as structural lumber, clear siding, cabinets or violins is different from liquidating entire old growth forests to make two-by-fours. In many instances, changing the scale of our activities would make it more obvious how we can achieve balanced forest use while maintaining meaningful employment.

As with other forms of ecologically responsible stewardship, we must learn to solve problems with finesse and ingenuity, rather than with more horsepower. Soft approaches that protect all the parts must be used rather than aggressive approaches that label some parts as valuable and other parts as worthless or harmful. Careful restoration would include five important principles:

1. Restore the forest at the stand level while making sure that these activities rebuild landscape connections.
2. Mimic natural processes.
3. Restore whole watersheds/large landscapes.
4. Prepare restoration plans and carry out restoration activities with local people, ideally those who inhabit the forest.
5. Treat the causes of degradation, not just the symptoms.

People experienced in agricultural restoration have found that degrading land use activities have often been designed by specialists and accomplished by powerful technologies, such as large machines and pesticides. In contrast, effective restoration requires all kinds of people with all kinds of skills. People with shovels will be as important as people with machines. Restoration must be more than a swift afterthought or hopeful solution to a single problem, no matter how commendable the impulse.

Some of the important activities that might comprise an adequate restoration plan include:

— Restoring soil health, including
— breaking up compacted soil surfaces
— introducing vegetation to stabilize soil, build soil nutrient levels, and restore water holding capacity
— Establishing natural drainage patterns
— Encouraging natural diversity by reseeding (instead of re-

planting) naturally occurring tree species
— Planting trees and shrubs where required for stabilization and diversification of a degraded forest community
— Carefully reintroducing animal and microorganism species
— Restoring riparian zones by reestablishing streamside vegetation
— Stabilizing stream banks and diversifying stream channels by reintroducing large logs (until natural large fallen trees are available)
— Carefully reintroducing natural—and sometimes human-induced—fire by limiting the practice of fire suppression to specified areas (near human dwellings and in wholistic timber management zones)

Towards Wholistic Forest Use

The single most important change in forest practices we must make has nothing to do with techniques or statistics or legislation. When I talk about changing forest practices, I am not talking about tinkering with the present system of timber management. I am not just talking about more parks and fewer clearcuts. I am not talking about changing political parties. I am talking about a change in our way of thinking.

As ordinary citizens, we must change our individual and community value systems and begin to accept responsibility for protecting the forest. Right now, both the forest and our communities bear the scars caused by our lack of creativity, responsibility and cross-pollination of ideas.

One of the promises of wholistic forest use is not only healthier forests but healthier communities as people come together to seek and develop a common ground. Those of us who are foresters or forest workers need to change, too. We must see our primary role as protection of forest ecosystems and advocacy for balanced forest use.

Most of all, we must change our way of thinking from a human-centered approach to an ecosystem-centered approach—from forest exploitation to forest protection, maintenance and restoration. The first step in this process is to begin to see ourselves as part of the forest ecosystem, not as an entity apart from the forest dedicated to making it more productive and predictable, redesigning it, manipulating and exploiting it.

Trees, the most obvious part of a forest, are critical structural members of a forest framework. But they are only a small portion of the structures, energy and organisms needed for a fully functioning forest. Humans can design a tree plantation, but it is not a forest. As more and more fully functioning forest ecosystems are converted to such plantations, we are losing more than big old trees—we are losing irreplaceable sources of pure water, habitat for native species, options for just settlement of land questions, the potential for sustainable employment and community economies, and opportunities for recreation and spiritual renewal—we are losing our means of survival.

Our forest practices in the future will need to begin from the humble understanding that the forest is an interconnected web that focuses on sustaining the whole, not on the production of any one part or commodity. All of the principles I have described—protecting and maintaining composition, structure and functioning, respecting biological limits, and limiting the scale—apply to every kind of human use, whether we are building a wilderness lodge, grazing cattle or cutting timber. Ecologically responsible practices accept the control of natural processes—nature at the wheel—and mimic the subtlety, diversity and unpredictability of natural changes, while using the forest carefully in a variety of ways. Wholistic forest use focuses on what to leave—fully functioning forests—not on what to take.

Of all the components of the forest web, the only one we know to be completely optional is human life. The forest sustains us, we do not sustain the forest.

Sources

Armstrong, Pat. 1992. "Canada, Brazil: No Comparison." Vancouver *Sun*, May 22, 1992

Bohn, Glenn. 1992. "Lumber Giant Targets Cashore." Vancouver *Sun*, ca June, 1992.

—"Native Forest Birds Dying Out, Federal Biologist Warns Loggers." Vancouver *Sun*, Friday, June 5, 1992.

Bowman, Colleen. 1991. Presentation to the BC Round Table on the Environment, Nelson, BC, June 4, 1991.

British Columbia, Forest Resources Commission. 1991. *The Future of Our Forests*. Victoria, BC.

British Columbia, Ministry of Forests and Lands. 1987. *Forest Tenures: Licenses and Permits for Harvesting Timber in British Columbia.* Victoria, BC.

—*Ministry of Forests Policy Manual* II- REC-003-1. 1982. Victoria, BC.

Forman, R.T.T. and Michel Godron. 1986. *Landscape Ecology.* Toronto: John Wiley & Sons.

Franklin, Jerry F. 1988. *Old Growth: Its Characteristics and Its Relationship to Pacific Northwest Forests.* Old Growth Conference, Corvallis, Oregon. August 25, 1988.

Gregory, S. and L. Ashkenas. 1990. *Riparian Management Guide: Willamette National Forest.* USDA Forest Service, Eugene, Oregon.

Grumbine, R. Edward. 1992. *Ghost bear: exploring the biodiversity crisis.* Washington, DC: Island Press.

Hammond, Herb. 1991. *Seeing the Forest Among the Trees.* Winlaw, BC: Polestar Press.

—1985. *Technical Evaluation of Forest Management Practices in the Nass River Valley.* Winlaw, BC: Silva Ecosystem Consultants.

Harris, L.D. 1984. *The Fragmented Forest.* Chicago, IL: University of Chicago Press.

Hunter, M.L., G.L Jacobson, Jr., and T. Webb, III. 1988. "Paleoecology and the Coarse-Filter Approach to Maintaining Biological Diversity." *Conservation Biology* 2(4), pp. 375-385.

Morrison, Michael L., Bruce G. Marcot, and R. William Mannan. 1992. *Wildlife-Habitat Relationships: Concepts and Applications.* Madison, WI: University of Wisconsin Press.

Plochmann, Richard. 1992. "The Forests of Central Europe: A Changing View." *Journal of Forestry*, 90 (6), 12.

Stott, Greg. 1990. "The Armageddon Option." *Equinox* 49, January/February, 52.

Thomas. J. Ward, C. Mazer, and J.E. Rodiek. 1979. "Riparian zones." In: J. Ward Thomas (Ed.), *Wildlife Habitats in Managed Forests.* USDA Forest Service, Agriculture Handbook No. 553, Washington, DC.

Aboriginal Forestry

The Role of the First Nations

Holly Nathan

THE SIGHT OF HAIDA ELDERS IN BUTTON BLANKETS defiantly blockading a remote Lyell Island logging road as RCMP officers read out sections of the criminal code is one of the more haunting images of BC's troubled relationship with aboriginal people. The fact is, the ownership of virtually the entire province has been under dispute for 120 years. Until agreements are reached with BC's aboriginal First Nations asserting their title to land, the question of who has control over BC's key forest sector will continue to overshadow everything to do with its management.

It must be stated at the outset that there is no one aboriginal viewpoint. From the half-dozen Vancouver Island reserves accessible only by float plane or water taxi to the urbanized Vancouver reserves of the Squamish and Musqueam Nations, BC's 196 bands and 180,000 aboriginals do not share geographic realities, histories or identical attitudes. Even the eleven distinct aboriginal languages are as various as Europe's. It is only possible to offer an overview, in the words of different spokesmen and spokeswomen.

However, many of the province's 415 native villages are struggling to overcome radical shifts in their way of life. Some have faced forced relocations to other reserves and economic dislocation. Most are dealing with astronomical unemployment rates and returning band members who have regained their native status.

With the forest economies located directly in their traditional territories, many communities are viewing the forest sector as a means to native economic self-sufficiency, and a way to break the shackles of a national $4.5 billion system of forced welfare and dependency on the Indian Act.

Of these, few are willing to participate in an industry that has caused so much damage without imposing their own standards. Almost without exception, First Nations leaders say they intend to blend tradition, cultural wisdom and a holistic attitude to the land in meeting their economic needs.

What settlements will mean to the forest industry or to the public is not yet clear. But leaders of the Nisga'a, the Gitksan Wet'suwet'en and the Nuu-chah-nulth have made their message known: there will be transfer of control over resources in some areas. There will be an aboriginal role in how resources are managed in other areas. And there will be a different approach to harvesting trees.

George Watts, chairman of the Nuu-chah-nulth tribal council, says aboriginal people will not agree to monetary compensation alone in negotiating modern-day treaties. He feels that, as with the signing of treaties for beads and blankets in the past, money will fail to provide any lasting guarantees for aboriginal peoples. They want a real say over land they claim as theirs: "Ten years from now, MacMillan Bloedel will be in for a shock. They will be coming to see me before they cut down trees." And such leaders as Kyuquot chief Richard Leo say that's in everybody's best interest.

In 1991, the NDP government agreed to negotiate land claims settlements and reversed a century-old BC policy of refusing to recognize aboriginal rights or title to land never surrendered through treaties. "It's unacceptable to perpetuate 125 years of injustice to native people," Premier Michael Harcourt said. Following the rejection of the Charlottetown Accord constitutional amendment, he promised to accelerate the process of settlement and self-government negotiations. At the same time, the Canadian government has vowed, perhaps rashly, that land claims will be resolved by the year 2000. In the meantime, BC's First Nations are effectively shut out of participation in the province's $11 billion forest industry. Virtually all the forest tenures in the province have already been allocated to major corporations and non-native interests. One planner with the Kwakiutl District Council on north Vancouver Island calls aboriginals prisoners in their own land. And report after report indicates the forest resources claimed by First Peoples are running out. Old growth on Vancouver

Island is destined to be eradicated by the year 2008, according to a 1992 federal government report on the national environment. The Nass River Valley, traditional territory to the Nisga'a, has been shorn. Mount Paxton, in the backyard of the Kyuquot people on the west coast of Vancouver Island, has achieved the dubious distinction of appearing in *National Geographic* as an example of clearcut devastation.

The Situation Now

In 1987, the Intertribal Forestry Association of BC(IFABC) was founded to address native concerns about forestry issues. In 1991, a native forestry task force travelled throughout the province gathering opinions on what aboriginal peoples wanted. Responses ranged from the need for job opportunities on BC government firefighting crews to demands for outright expulsion of major corporations operating in traditional lands and the takeover of control by native nations.

The association also collected some significant facts:
— Indian bands hold less than 0.5 percent of provincial forest tenures.
— Only one band, in northern BC, has a tree farm licence that allows it to engage in a long-term major business venture and comprehensive land management.
— A total of five bands hold forest licences.
— Twenty-two bands hold small woodlot licences to harvest wood.
— Only three BC aboriginals hold university degrees in forestry.
— A total of 1,200 natives are employed in the forest sector.

Moreover, IFABC found that forests on Indian reserves have been severely mismanaged. Under the Indian Act, which gives the federal government control over reserve land in a trust capacity for bands, forests suffered overcutting, inadequate tending and lack of reforestation at the hands of small contract companies and individual band members. The IFABC also found that aboriginal companies and workers received only eight per cent of the economic return from logging their own timber.

The association was advised that devastation to reserve land has been a breach of constitutional duty to aboriginal people and that damages could legally be awarded because of that neglect. New national native forestry legislation is now in the works and the federal government is preparing to transfer more control to native communities. A 1991 conference on native forestry in Nanaimo was told that if those forests were properly managed, they could generate a half-million dollars for native reserves.

Despite the IFABC's dismal findings, BC's First Nations are in a key position to turn the forest industry upside down. Unresolved land claim cases cover the whole of Crown land — and that comprises the total commercial forest base of BC. To date, twenty-two claims covering two-thirds of the province have been submitted to the federal government, with another eight or nine expected. The fact that the ownership of the land is under dispute has resulted in the loss of $1 billion in proposed mining and forestry investment as well as potential jobs in the forest sector, according to a 1990 report for the federal government by the accounting firm of Price Waterhouse. The Haida, the Ahousat, the Tla-o-qui-aht and the Lillooet have been at the forefront in the past decade of major confrontations and blockades, bringing logging operations to a halt and causing economic uncertainty. Companies like MacMillan Bloedel are in court rather than overseeing their tenures.

At the same time, BC is facing increasing international criticism for cutting down too much too fast and, on the coast, for hacking down one of the last pristine temperate rain forests in the world. Calls have come from environmentalists, from the public and, most recently, from the international community, for a major restructuring of the industry. BC's First Nations are the only group in the province with the legal, historical and moral clout to do exactly that.

Forces for Change

The Spiritual and Cultural Base
The culture of the First Peoples is imbued with images of the forest — as a source of spirituality, a place of refuge, a test of manhood and a place of cultural meaning. The concept of mass-scale use of the resource, with the wastage and devastation involved, is simply alien to the cultural ideology. Legends and ceremonies alone reflect a tradition of interconnection with the forest and of dependency on it.

Legends

Artist Tim Paul, a one-time resident carver at the Royal BC Museum, recalls one tradition. As he put the finishing touches on the cedar whale-man figure of a totem, he says that, according to his uncle Moses Smith of the Ehattesaht Nation on the west coast of Vancouver Island, prayers to each individual tree are offered before the trees are cut down.

In *Tales from the Longhouse by Indian Children of British Columbia*, a book published in 1973, the forest landscape is an integral part of the stories the children have heard from their grandparents and parents. In one, a young brave named Shan-Tec, or "boasting one," vows to climb the tallest mountain near Duncan on Vancouver Island. But he can only accomplish this feat through the kindness of a bear willing to help him cross a river and a deer generous enough to share berries with him.

In another, there are mysteries about the forest which may not be understood but must be respected. A great medicine man goes to the woods in search of the spirit of a child who has been stolen by an owl. He saves her spirit and returns to the village to dance and sing with the child's voice. But the miracle is conditional: no child is to go out under a full moon with an apple in her hand or the owl will take her spirit too.

In the tales of the birth of totem-pole carving and weaving, artistic and spiritual expression arise from contemplation in the forest. The man who becomes the first totem-pole carver wanders off across the waters to a forest clearing where he finds a dead cedar tree and begins shaping creatures from the wood. The young woman who finds a basket woven out of leaves has first roamed through the quiet forest to think about her mother and visit her grave.

Living Culture

The forests are not simply places of spirituality in legend alone. They are sources now of cultural renewal and reconnection to the land. The Malahat Nation, outside Victoria, is embroiled in a battle to stave off development on its sacred mountain, called Yoos or "Place of Thunder" by the elders. The South Island Development Corporation is planning a town—Bamberton—of potentially 15,000 people in an area currently used for Coast Salish initiation rites during bighouse winter spirit dancing ceremonies. These ancient rituals take place from November through to spring of every year on southern Vancouver Island, the Lower Mainland and in Washington State.

The continuing importance of aboriginal spiritual culture is reflected in the band's letter to Aboriginal Affairs Minister Andrew Petter:

> What the proponents fail to mention is that their development is proposed to occur on sacred native land which is already serving the traditional cultural, spiritual and material needs of the Malahat people. What the Bamberton proposal involves is the displacement of ongoing traditional aboriginal practices and cultural activities on our sacred mountain with a massive new $900 million development.
>
> We have many legends associated with Yoos. Historically, for anything our people needed from this mountain, a ten-day ritual was required . . . If Yoos felt the request was proper, it would be granted.
>
> Yoos continues to this day to provide for our people in many ways. It is a prime hunting ground, yielding necessary meat to see our people through the year. Our people continue to practice our aboriginal fisheries in the nearby waters . . . The forests and creeks of Yoos play an important part in the longhouse culture not only for our own Band but for all of the seventeen longhouses within the southern Vancouver Island area, where new dancers are taken for their bathing and other spiritual practices as part of the training and initiation ceremonies.

Near Prince George, Chief Peter Quaw of the Lheit Lit'en Nation defends another sacred mountain, threatened by logging in his area:

> Charlie Mountain is referred to as Grandfather Mountain, and is considered by the elders one of the most sacred areas in our tribal lands. Traditional chiefs and medicine people were the only ones allowed there to conduct sweats and other practices. The culture is reviving and we have to provide opportunities for these things. I'm 40. When I was 38 I had my vision quest. I was up on Charlie Mountain. It works. I had a speech impediment. I was punished at

residential school for speaking my language. I stuttered and stammered. But after I did my vision quest, I settled down. I talk better. I think better. I was up there two weeks.

The elders tell you what to expect, tell you things. They tell you whatever you take from the forest you must replace. In contemporary terms, you take it as sustainable development. Basically, that means every insect, every bird, every fish, every animal, every human being, has a right to occupy that land. A large number of our people go up Charlie Mountain within our territory.

Traditional Remedies and Other Cultural Skills

Another aspect of aboriginal culture being threatened is the medical knowledge and skills derived from the forest. In the Interior, one tribal group uses over 200 different species of plants for a broad range of ailments, yet current forest practices are eradicating traditional native plants.

For more than 20 years, University of Victoria professor Dr. Nancy Turner has been collecting information from native elders about the traditional aboriginal use of plants for food, medicine and culture. Turner, a nationally recognized expert on the topic, has documented aboriginal wisdom about using plants for everything from stomach ailments (tree bark) to fish nets and cakes (salal). A family remedy shared with Turner by Saanich elder Violet Williams, the "four-bark" remedy, involves a combination of barks collected from the "sunrise side" of various trees on southern Vancouver Island. They are boiled with licorice ferns, sweetened with trailing wild blackberry, and used as a general tonic for stomach pain, ulcers, tuberculosis and kidney problems.

Turner says, "The controversy over the western yew has brought this to the forefront. People realized the yew tree [which yields taxol] was in fact an important medicine in the treatment of cancer and that it was highly valued by the indigenous culture. But it was considered of no value to the forest industry."

Turner also says a recent study by University of BC graduate students shows that one of the most important medicines for the Carrier people in the Interior is derived from alder—another species considered of no value to the forest industry, and one that is actively killed off through the use of poisons or by girdling trees.

Forestry Experience

Traditionally BC's tribal nations are no strangers to the practice of forestry. Turn of the century photographs of remote villages like Fort Rupert near Port Hardy, home of the Kwakwaka-wakw people, show dozens of canoes pulled up on beaches studded with totems: a virtual shipbuilding industry in conjunction with the construction of homes, furniture and monuments.

Early coastal woods provided cedar logs and planks for houses, while woven rugs and room dividers were made from cedar bark and animal skin. Longhouses used wooden pegs for nails, and bark would be shredded, dried and used for clothing. Berries were picked from forest clearings, dried and elaborately prepared for potlatch feasts. And the construction of canoes ranged from vessels for everyday travel, to canoes for hunting, sealing and whaling, canoes for war, one-man canoes and huge, 60-foot seagoing vessels.

In addition, native involvement in commercial logging in BC dates back to the late 19th century, with individuals participating as loggers and running small, family-owned businesses, particularly on northern Vancouver Island and on the Haida Gwaii or Queen Charlotte Islands.

Gordon Antoine, chief of the Coldwater Band and then a leader of the Nicola Valley Tribal Council, told a symposium on Indian forestry that natives were highly skilled and knowledgable in the practice of forestry but have been forcibly excluded from participation:

> Up until 1949 we were not allowed the franchise and as such we were not citizens of this province. We couldn't go out and get timber sale licences or wood quotas and yet the sawmills on the reserves have existed since near the time of contact.
>
> The very management principles that we used to make sure we continued to have food and forest lands were outlawed. We couldn't start underburns anymore because the forest service said that we were destroying the trees. We were helping Mother Nature. We were shut out from the woodlots that went along with the agricultural land as part of the Pre-emption Act of this province. We were shut out of the quota system that allowed licence tenures in this province and the very transportation networks that we established by the event of haul licenses.
>
> That is the history that I know. A very viable economic system

was in place and was decimated through disease and an outright conspiracy to keep us out of the industry and all the industries in this province.

Rebuilding Through Forestry

Native communities are becoming increasingly involved in the forest economy—for the sake of their survival as a people. The tiny 160-member Ehattesaht band on northern Vancouver Island is one of the more controversial cases in point. Like many aboriginal communities that have been dislocated through development or economic upheaval, the Ehattesaht have no home.

For generations, family groups lived scattered on 13 reserves on the rugged west coast, their lifestyle based on an abundant fishery. But, with the decline of the fishery in the 1940s, the band's entire way of life changed. Employment had to be found elsewhere. Education became imperative. Today, the only place for children to go to school is in Campbell River, on the other side of the Island. Families relocate from the west coast to the east coast community every year to be with their children. There has been no place to gather to rebuild the community, to address language, education, and cultural needs, and to break cycles of social dysfunction. Even the band office is located in a mobile trailer on the side of a busy Campbell River road.

Ehattesaht chief Earl Smith's vision, guided by the elders of the band, is to create a new home at Zeballos, at the end of a west coast inlet. Already, 10 houses had been built by 1991, partly as a result of a joint venture initiated by the band in forestry operations. Another 12 are in the works. There are plans for a school, a learning centre and a gymnasium.

But the vision is partly dependent on how well Hecate Logging does. And Hecate Logging, a fifty-fifty partnership between the band and Port Alberni-based Coulson Forest Products, has hit headlines for exporting raw logs, for clearcutting in the Ououkinsh Valley (a name meaning Place of Many Wonders), and for carrying on its operations in the traditional territory of a neighboring band, the Kyuquot, without permission and against the Kyuquot's land ethic.

The Ehattesaht are one of only two bands in the province to hold a forest licence giving it access to major timber stands. Teaming up with Hecate three years earlier rescued the band from receivership, and the operation now generates $5.5 million in total revenue. "We are not against development," Smith said in a 1991 interview. "Without development, we have got nothing."

Devastating social problems for the Stuart-Trembleur band (now called the Tla'z'ten Nation) near Prince George, including high suicide rates and astronomical unemployment, resulted in an urgent push to compete for a Tree Farm Licence in 1981-82. It remains the only band in BC to hold a TFL, the longest term of tenure available under the Forest Act. Wholly owned by the band as a non-profit organization, Tanizul Timber operates through a half-dozen shareholders who take direction from the band council and membership. The enterprise currently employs 27 native staff and up to 100 for seasonal silvicultural work. Its activities include logging, hauling, log marketing and reforestation. While operations manager Victor Komori notes the company has played a role in improving the morale of the band members, the company does face difficulties:

> The amazing thing is, we are still here. And we are putting the emphasis on creating jobs. But it is difficult to mix business with social imperatives. We face high stumpage rates and high costs and, because we do not have a mill to process the wood as other corporate interests do, we have to pay during the down times to haul the logs to Prince George. There is not a lot of money left over for the band as a whole—it hasn't really benefitted. The government doesn't recognize us as a unique TFL holder, with unique problems. We have to log so much a year according to our licence documents. That causes problems in many of the villages which don't want logging because it goes against the traditional way of life. We are caught on the horns of a dilemma and we do the best we can.

Komori says it is only through land claim settlements that aboriginal people can hope to control royalties and logging operations on their lands to allow forestry operations to serve their interests.

The Tsou'ke (Sooke), just outside Victoria, are selectively logging the timber resources on their reserve land to provide jobs for members put out of work by the shutdown of a local mill in the late 1980s. The band is also building homes for women who are returning to the reserve after losing their Indian status when they married non-aboriginals and who then had their status reinstated in 1985. Chief Jacques Planes says the selective logging is being carried out with respect for the land and the resources.

One community that used its timber in a unique way to rebuild is the remote Ditidaht band on the west coast of Vancouver Island. Until last year, the band had one radio telephone and high unemployment rates. Now, it is talking about constructing new homes, bringing its scattered people back home and reestablishing links with its elders and its language. A telephone has been installed in the band office, a small tourist gas station and restaurant facility is being completed, there are plans in the works for joint park management and money has been put in trust for future generations.

Rather than protesting logging, the Ditidaht actually threatened to exercise its rights to cut down trees—and got a $9.6 million settlement from the federal government in 1992. The fact that eight of its sixteen reserves had been incorporated, without consultation, into the internationally renowned West Coast trail system in the 1970s was the crux of the issue. The government did not want to see the wilderness area damaged and tourism threatened.

In most cases, aboriginal forestry operations are intended to serve a collective social purpose, saving aboriginal languages on the verge of extinction and employing people who are being forced to shift from traditional trapping, hunting and fishing as a basis for their way of life. Employment for native people in rural areas also has spin-offs for the non-aboriginal local communities. Tanizul is estimated to have re-cycled $9.6 million into the local economy and is currently averaging $5 million in annual revenue. Such operations could be a key factor in the kind of restructuring that people like Stephen Owen, head of the Commission on Resources and Environment, are talking about when they emphasize decentralization, community involvement and accom-modation of social and environmental factors as essential to a new forestry regime.

Small Business

Until land claims are settled, and until some of the more innovative recommendations by the Intertribal Forestry Association (such as the creation of a new First Nations licence tenure) are adopted, native communities are attempting to gain access to the resource through the existing tenure system. The provincial government's Small Business Forest Enterprise Program (SBFEP) is a key to that involvement, however limited.

The program was a response to the 1976 Pearse Royal Commission report urging the government to move away from monopolistic corporate control by involving a mix of firms of varying size and providing opportunities for new entrants. Over 12 percent of the province's total annual allowable cut is available for sale to competitors through the program. About 2,800 individuals and firms are registered from the logging, small sawmilling, value-added and re-manufacturing sectors.

The government, however, has no idea how many native bands are involved in the program, partly because company names and joint ventures do not necessarily identify band affiliation. A study is now underway to determine native involvement. It is clear that native interest in SBFEP is high. It is this program, for example, that is providing the Toquaht Band near Ucluelet with the opportunity to be the first in BC to turn trees into a log-home business.

The tiny 70-member community has a 10-year timber sales licence and has established a small sawmill on their reserve. Three band members have been on a log-home construction training program in Prince George, and another three on a sawmill training program. Band spokesman Gary Johnson said in 1991 that the Toquaht believe its business is viable, despite layoffs earlier that year of half the work force at MacMillan Bloedel's Kennedy Lake division at Ucluelet.

"We are hoping that, because we are doing processing on site in a smaller-scale operation, we can find a niche market for our products," Johnson said. "There is lots of competition and we are a new company starting out. But there are opportunities in the Tofino-Ucluelet area for new homes."

By early 1992, only two bands in BC had timber sales licences for value-added products, the Toquaht and a Burns Lake band. Of 500 applications received by SBFEP for initiatives in business from furniture manufacturing to home construction, only three were known to be from native bands. A consultant's review of the program showed it falls short of addressing two major concerns aboriginal leaders have expressed about their involvement in the resource: it has been criticized for failing to emphasize strongly enough remanufacturing and value-added initiatives, and for not promoting improved forest management.

Small Woodlots

The woodlot program, attracting the involvement of 22 bands to date, was initiated in 1979. It provides a licence to individuals or organizations who qualify to manage small areas of Crown land, sometimes in conjunction with private land or reserve land owned by the licensee. There are about 460 woodlot licences currently in existence. In 1988, four percent of the cut in the program was allocated to native bands.

A review of the program by consultant David Gillespie in 1991 noted the industry was under siege on the question of forest management: "The popular topics such as clear-cutting, overcutting, pollution and land claims, all led to a public outcry for better forest management." He said the woodlot program was a move toward the goal of providing smaller, community-based operations. Bands wanting to become involved in integrated resource management of their reserve lands (taking into account forestry, wildlife, water and other ecological factors) have applied for the smaller licences.

From the native perspective, however, the problem is that the licences are too small—400 hectares is the maximum—to provide any degree of job creation or economic development. And the two dozen woodlot licences "probably cost the bands more in administration than they generate in profit," according to the Intertribal Forestry Association.

Innovative Programs

The Lheit Lit'en Nation near Prince George—angry that its sacred territories were threatened in an area so extensively clearcut it can be seen from the moon—applied to manage 200,000 hectares of old growth in the Herrick Valley under a $54 million, six-year Model Forests project initiated by the federal government in 1991.

The application was unsuccessful, but it illustrates the interest of many bands in implementing more sensitive forest management techniques. The Lheit Lit'en Nation, in conjunction with a local environment group, wanted to eliminate clearcutting by Northwood Pulp Company (Noranda), which planned to harvest 4,500 hectares over five years. With other small businesses active in the area, 50 per cent of the wilderness was expected to be affected in that timeframe.

The First Nation proposal was to use an integrated resource management scheme employing five to ten times more people by training natives and others to use horse logging, selective logging and small-machine logging techniques in the watershed. It also proposed more intensive small-patch logging (no larger than 15 hectares) in appropriate areas. The Lheit Lit'en plan attempted to balance the harvest of trees for commercial use, the interests of other user groups, and the ecological characteristics of the area to create a stable, vibrant local community.

Joint Ventures

The only other means aboriginal communities have of acquiring more timber-cutting and forest management rights through the provincial tenure system is to buy out non-native companies that hold tenures or to form joint ventures with existing companies.

Ainsworth Timber is one of the last family-owned operations in BC. It employs about 100 people and owns five mills. The company decided to work co-operatively with the aboriginal community after three bands in the area threatened to blockade a proposed bridge unless they were allowed to take part in the economic development in their traditional territory. The company leases reserve land from the Fountain Lake band (Xax'lip Nation) for its Lillooet mill and, at 100 Mile House, has silviculture contracts with the local Canim Lake and Dog Creek bands.

Kelly McCloskey, Ainsworth's chief forester, says, "We pay the natives for the right to operate the mill, and the income earned by the band members who work there, making up half the work force, is tax exempt. There is an automatic mutual benefit. We give the bands first rights [on silviculture contracts] because the Ainsworth family has lived in the area since 1952 and sees the bands as part of the community. They see the resolution to the current debate is to bring natives into the economic community. That is what most of the bands want. The bottom line is not necessarily the sole object of all companies."

Dave Archie of the Canoe Creek band described the process to an Intertribal Forestry Association hearing as follows:

> We want to incorporate the Indian philosophy into clearcutting and we have discussed it with Ainsworth. We said, "Before you make a

desert out of our backyard, can we talk to you? They said okay."

We want to be a part of the process since this is our traditional area, we have lived here for the last two to three thousand years. The land is our meat market, fish market and we have trainees in the fish and wildlife fields to look after the lakes and restock them. That's where we are coming from. We are tired of being a welfare state within the federal and provincial systems.

Tony Shebbeare, of the Council of Forest Industries, said in a 1991 interview that most companies have shown a change in attitude about relations with aboriginal communities over the past five years. Major corporations like MacMillan Bloedel emphasize they are hiring more aboriginal people within their corporate structure. "We are not stupid," said Shebbeare. "We can see where things are going. One way or another, [First Nations] will have a bigger role than they have now. To prepare all of us for that, we have to start working together. Natives will be part of a planning team. But it won't be totally equal partnerships. I don't see that for quite some time."

Doman Industries on Vancouver Island is taking a more pro-active stand with the Kyuquot, consulting the band on how the company's forest licence operations will be carried out in traditional territory claimed by the band and not yet settled through negotiations. The company says it can balance its profit interests with the band's environmental concerns, even if that means slowing its rate of cut, avoiding clearcuts or changing its management priorities.

According to the company's chief forester, John Hawthorne, "Our objective is to ensure we don't end up with blockades. We want to be at the leading edge of change, not the trailing edge."

Chief Richard Leo says the Kyuquot supported the Meares Island blockades in the 1980s to stop MacMillan Bloedel's operations and has worked with the environmental and non-native Kyuquot community to address the devastating impacts of logging. But three years ago, it decided to meet with Doman: "To call our area the Brazil of the North is not too far-fetched. We could complain and be bitter about what's happening. But clearly the issue is getting it fixed. We have always stated we were never against logging. But we asked for a change because we didn't want the Tashish watershed trashed because of clearcut logging."

The problem with joint ventures between industry—which can provide the capital, the business acumen and management skills—and First Nations—which can supply land, resources, labor and economic incentives through tax arrangements and other financial advantages on their reserves—is that economic imperatives are more likely to be emphasized than political or environmental concerns. Gary Merkle, one of BC's few native professional foresters, warns there are dangers for native nations in joint ventures, though he is generally hopeful:

> Some people think they are compromising their positions by involving themselves with the same companies responsible for the devastation at their back doors. In the case of small and often isolated bands, they can be used and abused by manipulative corporate interests.
>
> But I look at it like this: we have a philosophy built for the Tahltan territory. If you are well-enough grounded in your principles and can cut your business loose from the partnership at any time, you can use the system to help you find the things you need when the day comes and you sign some kind of treaty.

Tourism

Tourism is another major goal of aboriginal communities that could potentially meet their economic needs, while protecting the resource for traditional purposes. Native bands are aware of their unique positions in some of the world's last wilderness areas and want to play a role in BC's $4.8 billion tourism industry.

The newly formed First Nations Tourism Association estimates tourism could generate $400 million for BC bands and tribal councils through, among other things, the establishment of gas stations and restaurants on native land, guided tours and traditional native events. The association estimates there are already 400 to 500 tourist-related band activities in BC, but to date no effort has been made at coordinating the native cultural tourist industry.

The Future

Government and Aboriginal Relations

Native anxiety about the fate of the forests left on their traditional territories is heightening in a climate of rapid political change as land claims negotiations appear closer to reality.

The federal, provincial and First Nations representatives of the BC Land Claims Task Force, which drew up the 1991 blueprint for settlements in BC, recognized that conflicts would continue until treaties are signed. They unanimously endorsed a recommendation that measures be taken in the interim to protect First Nations interests where they could be affected by development and resource exploitation. Aboriginal interests do not have any meaningful protection otherwise under the current legislative framework. "Interim measures agreements are an important early indicator of the sincerity and commitment of the parties to the negotiation of treaties," the task force report pointed out.

The BC government, for its part, has concentrated its efforts in the first year in office on "joint stewardship" agreements with bands to "bridge the gap between the status quo and the new regime," as Petter explained, nearly a year into the NDP's first term in office. The question of moratoriums on logging operations or reallocations of resources will wait for comprehensive claim negotiations. Petter said, "We will improve consultation with aboriginal people. There may be some opportunities to enhance economic opportunities through silviculture, road maintenance, improved social programs. But the actual transference of the resources, or moratoriums, should be kept for the treaty negotiation table. We are not about to shut down the entire province, or shut down huge areas, in advance of those negotiations."

Three bridging agreements have been reached, not without controversy.

The agreement with the Fountain Lake or Xax'lip First Nation promised an increase in their involvement in land and resource use in their traditional territory, based on the NDP's recognition of aboriginal title and inherent right to self-government. The Cowichan band near Duncan signed an agreement with the Environment Ministry to consult on issues of concern in key watersheds. And an agreement with the Haida promises joint management of the recreational fishery.

However, already there is a sense of betrayal. Don Ryan, speaker of the Office of Hereditary Chiefs of the Western Gitksan in northwestern BC, charges that the BC government could have proven its willingness to settle land claims and reallocate resources by transferring the major forestry licence belonging to Westshore Terminals Ltd. to the band when the company ran into economic trouble in 1992. Instead, it turned it over to Repap Carnaby Ltd.

It was the Gitksan who charged Forests Minister Dan Miller, an employee of Repap, with conflict of interest in making his decision, a charge resulting in his three-month suspension in September of 1992. Ryan said he was skeptical about the "empty promises" of the federal and provincial governments to resolve land claims, accused the province of lacking coherent policies and political will, and charged it with paying more attention to the interest of third parties, such as MacMillan Bloedel and Repap, in determining its land claims policies than to aboriginal rights.

Miles Richardson, chief of the Council of the Haida Nation in the Haida Gwaii, has also indicated that First Nations concerns about disappearing resources in their territories are not being treated seriously in the interim.

There are instances of disillusionment at local levels of government, too. The Quatsino, a 267-member Coal Harbor band near Port Hardy, want to preserve two of the last six remaining watersheds on Vancouver Island for research purposes, tourism and protection of traditional values. They have become frustrated with what they say is a merely "token" role on the Brooks Planning Advisory Committee, a local advisory board providing input on resource management decisions in their area. They say their concerns about logging in the Klaskish/East Creek were being ignored despite support from the Ministry of Environment, which was equally concerned about environmental issues such as wildlife and old growth in the watershed. The band withdrew its representative in 1991.

"I don't want to be involved in a committee that will rubber stamp logging plans," Quatsino Band councillor Ralph Wallas said. "If my grandchildren and yours want to know what watersheds used to be like on Vancouver Island, this will be one of the last areas for study."

One of the landmarks in aboriginal-government relations that is likely to be influential in the future is the Gitksan land claim case. The controversial BC Supreme Court decision by Chief Justice Allen McEachern ruled against recognition of the Gitksan Wet'suwet'en claim to 57,000 square kilometres in northwestern BC (a decision now under appeal) after a three-year, $25 million court battle. The ruling did, however, result in new obligations for provincial government that internal government documents have indicated could compel the province to favor aboriginal interests over logging, mining and other development interests in the province. Orenda Forest Products Ltd. received approval in principle from the BC government in 1992 to build a $400 million mill between Terrace and Kitimat on condition that it consult with the region's native bands.

Meanwhile, the fate of the province's land base is being determined not only on the land claims front, the legal front and the constitutional front, but through the establishment of the Commission on Resources and Environment.

Commissioner Stephen Owen has stated that his work in drawing up a comprehensive land use plan for the entire province will incorporate and support the results of the treaty-making process with First Nations. "We have to see the issue of land use and environmental protection as operating within a climate of very rapid change in terms of aboriginal participation, rights and justice in this country. And forest companies aren't interested in who owns the land or resources, but in the control of the timber flow, in economic certainty and stability."

The fact is at the moment, Owen said, the resource economy is suffering: "Tens of thousands of jobs have been lost over the past couple of decades. One-industry towns are disappearing or are in desperate situations. Large industrial interests are leaving the province." But he said in an interview that he thinks the industry can be fundamentally restructured to bring about economic and social health:

> We are drawing on traditional native culture to rethink the way we have been exploiting resources in western culture. An essential, highly sensitive and complex responsibility of the commission is to ensure effective liaison and co-ordination with the self-government and land claims negotiations involving First Nations. It recognizes stewardship, ownership, management and users of land areas could change to native control. And CORE wants to develop a process that would have the power to look at moratoriums, deferred logging activities, redirection to other less sensitive areas.

The general public is clearly disgusted with confrontations and interest groups are becoming exhausted by it . . . I think if we are creative and generous in our thinking, we can work out a set of options for First Nations.

The Environmentalist Alliance

Since the blockades of the 1980s, aboriginals in BC have been associated with stopping forest operations. But their message is more complex than saving trees. It's about aboriginal control over resources. And that has led and will continue to lead to a somewhat uneasy relationship between aboriginals and environmentalists.

In the September 22, 1992, edition of *Ha-Shilth-Sa*, a Port Alberni-based tribal council newspaper, the Ahousat and Tlo-a-qui-aht First Nations distanced themselves from environmental protests in Clayoquot Sound. The two bands expressed concern that recent protests were becoming violent and encouraging acts of vandalism against local logging companies and their personnel. They also expressed dismay that their territory had not been recognized or their permission sought in the protest activity. The two First Nations took the opportunity to note they are currently involved in negotiations with the BC government to develop a resource management scheme for Wah-nah-juss/Hilth-hoo-iss or Meares Island.

It may have seemed a surprising move in light of the alliance between aboriginals and environmentalists that successfully stopped logging on Meares Island in 1985. But the Tla-o-qui-aht Nation is not anti-logging. More than 60 per cent of the band's 600 members used to work in the forest industry before the 1985 injunction stopped MacMillan Bloedel from carrying out its operations. Chief Francis Frank said the band was angry because there was going to be no resources left on traditional territory once land claims were settled.

"Our people are not naïve," he said. "We are changing with the times along with everyone else. And that means some economic development and logging. There are definitely areas we want to preserve. But we don't support the idea of complete preservation. There is a need for consultation on all sides."

In fact, native groups have ended up the target of environmental outrage. A coalition of environmentalists picketed the offices of Coulson Forest Products in Port Alberni in January of 1992 because its joint venture operation with the Ehattesaht Band involved clearcutting a north Island watershed and selling raw logs to Japan.

Another recent controversy, at Stewart in northern BC, cast the Kincolith Band in the role of environmental threat. The town was in an uproar about the band's plans to log on a reserve visible from the town. But the band council, located 100 miles away, had in fact been negotiating for years, first with the Social Credit government and then with the NDP, to exchange that stand of timber for another, less controversial one closer to its reserves. The 400-member band is facing high unemployment, with 195 applications for any kind of job that becomes available.

Kincolith member Harry Moore says: "It's very depressing. Our councillors and chief have been on our case and told us to do something now. The situation is creating a lot of hardship. And we have a whole slough of experienced loggers who used to work on the outside. I was a logger myself."

The band is looking to a joint venture with Logics Enterprise to help it address its needs: "This is the first big project and it could turn our lives around. We have tried logging in the area before and they told us they don't want to see the area scarred by logging operations. But there are a lot of scars all over BC. And exporting raw logs is a possibility. But we are trying to be as fair as we can so we don't hurt anyone."

Differences between the two interest groups were set aside recently when several prominent aboriginal leaders were asked to speak at a conference of the BC Environmental Network (BCEN), representing over 150 environmental organizations. Until then, representation by native people in the BCEN had been minimal. Mike Timney, Green Party spokesman, said environmentalists sometimes worry that native leaders will promote economic development over environmental protection. For their part, native leaders feel environmentalists don't understand their communities' need to create jobs and economic opportunities as well as protect the environment.

In at least one instance, however, a new alliance is developing. In June of 1992, a coalition of First Nations, labor and environmentalists continued work on developing consensus on a new Forest Stewardship Act—a movement begun in 1990 out of a sense that the forests had been failed by government, laws and institutions. The Tin-Wis Forest Stewardship Act is an attempt to establish sustainable community control over forestry operations in the hope of reversing trends toward massive layoffs and falling forest revenues. Tin-Wis is supported by the David Suzuki Foundation, the Canadian Pulpworkers Union, the Sierra Club and the Nuu-chah-nulth tribal council in Port Alberni under George Watts.

Land Claims

The NDP government in 1992 acted on one recommendation of IFABC by creating a 12-member First Nations Forestry Council mandated to increase:

— Joint ventures in all aspects of forestry.
— Aboriginal participation in each forest district and region.
— Aboriginal involvement in silviculture.
— Access to tenures where opportunities exist.
— Training and education.
— Participation in planning and in protection of First Nations heritage sites.
— Jobs for aboriginal people in government and industry.

But it is clear this is not going to be enough. In fact, most aboriginal leaders say co-management and joint stewardship are no substitute for aboriginal title and rights. Nor will land claims settlements without economic benefit be satisfactory.

Meanwhile, a February 1992 Southam poll testing national popular opinion on native self-government showed support of the principle but hesitancy about the details. The question of what kinds of arrangements should be made regarding natural resources on native lands showed that 43 percent might support native land ownership with the proviso that royalties be paid to federal and provincial governments, 17 percent favoured outright ownership, in which First Nations would keep all revenues, and 29 per cent preferred to see royalties paid to natives for resources used on their land and ownership remain unrecognized.

Fears surrounding land claims settlements include the idea that major tracts of land will be transferred to native control, that forest resources could be less effectively utilized, that the public will lose royalties, that regulations will become bureaucratic and complicated, that native-controlled enterprises would not be able to cope with a complex world timber market, and so on. On top of this is the undeniable fact that BC's largest industry is in decline—$10.9 billion in industry sales in 1991 accounted for only 13 percent of the total provincial output, down from 20 percent five years earlier.

The question is how these fears and competing interests are going to be addressed, and what aboriginal land claim settlements are going to mean. There are some indications. As we have seen, native bands are not totally preservationist—many have been employed or involved in the forest sector for generations. Some bands are involved in logging on their own reserves and are pushing for better management practices. Some emphasize job creation in commercial ventures, others emphasize a management role in resource use. Still others are at the forefront of a community-based approach to forest management with an emphasis on social benefits. Native corporations have been formed to conduct logging operations, or sawmill businesses, or to join hands with business partners in commercially managing the resource.

One study, *After Native Land Claims*, examines various initiatives already undertaken by native communities in the province as bellwethers for aboriginal involvement in a post-land claim BC. In the final analysis, it takes a positive view that the aboriginal role will provide both economic and conservationist advances.

In any case, change towards greater native control and management of forestry is on the way. In a few words to a small gathering on a hot spring afternoon in 1992, Willy Seymour, a spokesman for the 54 nations of the Coast Salish people, captured both the loss and the hope involved in aboriginal forestry's future. Speaking to the Royal Commission on Aboriginal Peoples, during its visit to a traditional West Coast longhouse near Victoria, Seymour said:

Our people were the first loggers. The posts in this house, they are sacred—for our shelter, our government, our education, for our discipline and justice.

Our forefathers selected the trees that would carry the echoes of our people and our history. Our forefathers used to go to the beaches and gather food; today that is no longer possible. They used to hunt deer and gather roots from these hills. That is no longer possible.

We were the only users of the resources. Our relatives were the centre of our empire where we went to practice our spirituality. The teachings in this home never change, nor the ultimate respect we hold for the land.

To ensure that spiritual values are protected, we need the land base. If we continue to use the resources, we must contribute to their enhancement. Today I feel a glimmer of hope that the commission has come into our home to listen to our concerns. I hope you will take the heartbeat of my ancestors with you.

Voices of the Peoples

On Logging

Bert Mack is the chief of the tiny Toquaht band near Ucluelet on the west coast of Vancouver Island. He worked 40 years as a logger in the industry. He and his four brothers figure they have 170 man-years of logging experience between them working for MacMillan Bloedel on lands their ancestors once considered theirs. In a March 1992 interview, he said:

> Our people have never benefitted. We watched what was happening and we didn't agree with it. We are not anti-logging. But we figure there are proper ways of doing it.

On Social Needs

Tribal chief Justa Monk, of the Carrier Sekani tribal council, spoke at a 1990 meeting in Prince George about the imperative of resolving land claims:

> Natives were herded together, then moved to small, resource poor reserves. There they attempted to live traditionally but because of the small land base they were unsuccessful. They tried to join the logging and mining work forces but they were shut out and left unemployed. Reserves soon became, not communities of proud, independent people, but communities of dysfunctional families, many relying on welfare to feed and clothe themselves, and on alcohol to relieve the boredom.
>
> If and when the land claims issue is settled, native people will once again have control of their traditional territories. Former rela-

tionships between corporations and government regarding rights to exploit forest, mineral and water resources will need to be changed. Another party, Indian First Nations, will be sitting at the table to discuss the use of the resources located in their territories.

I know it is impossible to turn back the clock. We may never have a chance to live off the land again, but we do have a chance to participate in the economic activities based on our natural resources so that we can make a better future for our people to have a decent life without drugs and alcohol. To make a native business work you can't run it on 50,000 cubic metres to any one band. We want co-management of the resource.

On Native Exclusion from the Industry

Gordon Antoine of the Coldwater Band spoke at a symposium on Indian Forestry:

I want to stress that even before contact, Indian people practised forestry and traded in the products of the forest land. It is not something new brought by the white man. Our language, and I believe just about every language in this province, has a word for lumber, a word for tree. There is a name for every kind of tree in this province in our languages. That was not invented by the new structures governing this province. The trade routes that we used were the very routes now taken over and renamed. One of them is even being talked about as a historic park site. We traded in the products of wood. My interior people traded with the coastal people for cedar for canoes. One cedar canoe was worth one slave. We traded in medicine made from the products of our forest land, yet even now people are still trying to develop more commercial alternatives to natural products like the sap from the balsam tree that we have always used. Historically, we have always managed our forest lands. The whole business of underburning that is now beginning to be rediscovered by the BC Forest Service was practised by our old people on a fairly regular basis so that the various foods that we eat, the various foods that the animals eat would come back year after year in our forests. Forest management is not new. It is built into our history. Commerce is not something new. Commerce is built into our language. The very processes of trading, there are words for these. A trading dialect was established among us.

I have found evidence of this on Vancouver Island as well as in the north-east corner of this province.

Dean Dokkie, Headman, Treaty 8 Tribal Association, said in Fort St. John in November 1990:

> The situation with Louisiana Pacific has been a major concern to the three Bands concerned: West Moberly Band, Salteau Band and Half-way Band. The company has been taking out the spruce and pine for the last 30 years right in the middle of the Bands' hunting and trapping territories without any consultation with the Native communities affected. There has been no consideration of the impact on the communities and no consultation involved with any planning or decisions affecting the areas.

On a Holistic Approach to Forests

Tom Sampson, chairman of the First Nations of South Island Tribal Council, spoke at an Intertribal Forestry Association hearing:

> We are trying to organize ourselves to deal with the forest and we are not talking just about trees, it is more than just trees. Everything that is in the forest is all one and the same thing, when you talk about forest you talk about everything that is in it. That includes mining, hunting, trapping, gathering of foods, medicine, and the list goes on.

Chief Roger Jimmie of the Kluskus Band said:

> In our case there is a holistic view of forest management and we do not look at it in the sense of a timber block but more as a way to protect the whole watershed. Not only that but it will protect the berry patches, it will protect hunting and fishing and it will also help to prevent water erosion, wind erosion, you name it.

Cameron Beck of the Chilcotin-Ulkatcho-Kluskus Nations Tribal Council said:

> Industry people laugh about the importance of the mushrooms to the Native economy. The Ministry needs to recognize the importance of

the mushroom ecosystem, caribou habitat, tourism values and other things valued by the Native Communities. The Native people learn more if the forest is the school rather than the school being a classroom.

On Native Participation in Development

Pat Edzerza, chief of the Tahltan Nation in northern BC said:

> In our traditional territories we have the largest block of traplines in BC, which is an important source of income for our people. These areas are being logged and our people were promised jobs to replace trapping but they haven't happened. We are not opposed to development. We want our people to develop along with the resources. Our traditional way of life is being taken away and we want it replaced with something our people can take to the bank. Social problems have come with development and we need programs to help our people.

On Labour and the Limitations of Make-work Programs

Richard Watts, co-chairman of the Nuu-cha-nulth tribal council in Port Alberni, said in a 1991 interview that his tribal council had developed a tree-seedling nursery and was selling its products to MacMillan Bloedel and Fletcher Challenge. But he called business a "dog-eat-dog world that native people are not all that attuned to." He also said creation of "aboriginal licences" to allow native communities long-term access to timber is one idea that should be considered:

> We end up with small licences and two months of work a year for a whole tribe. That is not the kind of opportunity we are looking for. Our people's attitudes have to change too. We have to show people we are willing to go out and do the jobs.

On the Native Relationship with Unions

Chief Russel Kwasistala of the Kwakiutl District Council expressed concern:

> Another thing is that we were knocked out of the forest industry because of the IWA union. Many of my people on the coast were seasonal loggers. They logged until fishing season started and went

back to it after fishing was over. Our traditional way of life was seasonal and nomadic and doesn't fit in with union seniority.

On Business Difficulties
Chief Pat Alfred of the Nimpkish Band at Alert Bay on Vancouver Island said:

> We have had successful students in Alert Bay in the forestry program there. They got the training but there was no work for them when they finished. The Bands aren't funded to make economic development ventures like forestry. Why is it that Kingcome Village sits across the river from a logging camp but only one person from the village out of 200 people is employed in logging?

Chief Earl Smith of the 160-member Ehattesaht band on the west coast of Vancouver Island, said his band was one of the first in the province to acquire a forest licence in 1972 and has had a stormy business history in managing the operation. One of the hardest lessons concerned the issue of mixing business with social programs. One of the stipulations of a $2.1 million fund from the federal government to help with the operation was that native people be hired:

> The condition was imposed by the government to create jobs. Very few of our people have been involved at the business level in the forest industry. We had social problems—the use and abuse of drugs and alcohol. The question is not whether to hire native people or not, it's whether you can make the business work. Every tribe should go through what we went through. They could understand then: you have to be chewed up and spit out a few times before you can get into business.

On the Need for a Different Kind of Long-Term Tenure
Cameron Beck, Chilcotin-Ulkatcho-Kluskus Nations Tribal Council, said at a hearing in the Cariboo area:

> The Kluskus and Ulkatcho Indian Bands applied for wholistic TFL's three years ago. The question arises, why tenures? There are three

basic reasons: financing, management and social and education concerns and opportunities. On the issue of financing it is much easier to go to the bank and arrange financing if you have a tenure than if you don't. As for management, the Ulkatcho want to protect things like tourism opportunities, caribou habitat, and mushroom ecosystems. They have a much better chance of doing this if they have tenure. People are much more interested in working on a long term basis and being responsible if they can feel they have a stake in the land and resources. As for the social, educational and economic opportunities, the Bands call these Wholistic TFL's which means they are more concerned about issues such as the caribou habitat etc. than is evident in other forms of forest management. We want the opportunity to teach and train the people of our communities towards a goal of long term employment.

On the Need for a Real Voice

George Harris of the Chemainus tribe and mid-Island tribal council spoke during a January 1991 Intertribal Forestry Association hearing:

It is clear our people need jobs, but unless our Elders and the Indian People are involved in forest management decisions at the top level there will be no jobs left for anybody, let alone Native People.

In the state of Washington, native people are selected from their tribes to the State Forestry Management Board where they have direct decision-making with full veto power. In BC far too often native people have been given token positions on advisory boards with no real commitment from this province to give natives full management decision-making powers. Right now there is total disagreement between the powers about how to manage the forests. Native people are at the table willing to negotiate and we have yet to be listened to.

Major Land Claim Disputes as of November 1992

Meares Island

A court case scheduled to last 120 days was adjourned for the second time in November 1992 in order to negotiate a settlement to the dispute. The Ahousat and Tla-o-qui-aht (formerly Clayoquot) won an injunction in 1985 preventing MacMillan Bloedel from logging timber it claims is worth $30 million. The trial is scheduled to resume August 31, 1993, if settlement cannot be reached. The two First Nations are claiming a right to exclusive use of the trees on Meares, located near Tofino on the west coast of Vancouver Island. Current negotiations involve a greater aboriginal role in resource use on the island, from trees to water rights.

Deer Island

The dispute involving the Kwakiutl band of Fort Rupert near Port Hardy on northern Vancouver Island involves an island which lost its reserve status some time between 1851 and 1932. The band is claiming the island as part of a specific claim for cut-off reserve land, and as part of a comprehensive claim for traditional territorial rights. Private contractor Halcan Log Services Ltd. bought the island in 1986 from MacMillan Bloedel. Protests by the band brought logging to a stop in 1986. The company's attempt to win an injunction against the occupation of the island was denied by the Supreme Court of BC, which ruled logging would be pre-emptive of the Kwakiutl land claim. The issue has not gone to trial, but an injunction remains in place. In the meantime, the band is involved in negotiations with the company and the province to arrive at an out-of-court settlement.

Delgamuukw

The claim by the Gitksan Wet'suwet'en to 57,000 square kilometres of territory in northwestern BC was quashed in a controversial decision by Chief Justice Allan McEachern in BC Supreme Court in March 1991. He ruled that aboriginal title had been lawfully extinguished by the Crown during the colonial period. The three-year, $25 million land claims case was the longest in BC history. An appeal decision in June 1993 held that the Gitksan Wet'suwet'en people had "unextinguished non-exclusive property rights" to the land.

Nemiah Valley Indian Band

A 1991 injunction against Carrier Lumber Ltd. of Prince George prevents logging operations in the Chilcotin that threaten two major traplines owned by the band, one around Chilco Lake and the other at

Taseko Lake. The injunction is still in place and no negotiations have yet been undertaken.

The Pasco Interlocutory Injunction

This involves a dispute over planned double-tracking of the Canadian National Railroad in the Fraser Canyon near Ashcroft. The company's plans were protested by dozens of bands in the area on the basis of interference with their fishing rights. The injunction was served in 1985. No trial has yet been held.

The McLeod Lake Indian Band

This band, near Prince George, won a BC Supreme Court injunction in 1988 against a number of logging companies and the Ministry of Forests, preventing logging activity on 55,000 acres of Crown land claimed by the band. The band falls within Treaty 8 boundaries and argues it is entitled to treaty rights. The injunction will remain in place until such time as negotiations or the courts determine the extent of aboriginal rights. The 380-member band has in the meantime formed a community-based committee to work toward an alternative solution to litigation, and to achieve increased native participation in the industry.

Haida Gwaii (Queen Charlotte Islands)

A 1990 injunction involved 80 acres of land near Masset where a high density of culturally modified trees is located, and included evidence of bark-stripping, campsites and canoe planks. The dispute resulted in an agreement to allow Stejack Logging Ltd. to harvest only in specified areas. But the conflict has sparked a joint effort by the Ministry and the Haida to create an inventory of culturally modified trees on Haida Gwaii. The project is scheduled to be completed by mid- 1993. In the meantime, the Ministry is deferring harvesting in culturally sensitive areas in favour of less controversial territory unless an agreement can be reached with the Haida.

Sources

British Columbia, Ministry of Aboriginal Affairs. 1992. *The Aboriginal Peoples of British Columbia: A Profile*. Province of British Columbia.

—1992. "Province and Xa'lip First Nation Sign Joint Stewardship and Other Agreements." News Release, 6 July.

British Columbia, Ministry of Forests. 1992. "Province to Establish First Nations Forestry Council." 18 August.

British Columbia, Ministry of Environment, Lands and Parks. 1991. *Co-operative Management Agreements: A Literature Review.* Occasional Paper No. 1. June.

—1992. "Cowichan Joint Stewardship Agreement Signed." 30 June.

British Columbia, Ministry of Tourism and Ministry Responsible for Culture. 1992. "Province Signs Joint-Stewardship with Haida." 28 July.

Cassidy, Frank and N. Dale. 1988. *After Native Land Claims? The Implications of Comprehensive Settlements for Natural Resources in British Columbia.* Oolichan Books and the Institute for Research on Public Policy.

Commission on Resources and Environment. 1992. *Report on a Land Use Strategy for* BC. Government of British Columbia.

Forestry Canada. 1991. *Model Forests: Background Information and Guidelines for Applicants.*

Gillespie, David, W. *Small Business Forest Enterprise Program. Woodlot Program. Program Review.* 1991. Commissioned by Minister of Forests Claude Richmond in Sept. 1990. Report to Minister of Forests.

Indian Children of British Columbia. *Tales from the Longhouse.* 1973. Sidney, BC: Gray's Publishing Ltd., BC Indian Arts Society.

Intertribal Forestry Association of BC. 1987. *Indian Forestry—What does It Mean. Proceedings of the Symposium,* Oct. 28-29, 1987. Shuswap Cultural Centre, Kamloops, BC.

—1991. "Lands, Revenues and Trusts Forestry Review." *Report.*

Lheit-Lit'en Nation. 1992. *Co-operative Management of the Herrick Valley Old Growth Forest and Aquatic Resources. An Opportunity Under the Federal Government's Initiatives on Developing a "Model Forest."* A proposal submitted to Forest Canada's Model Forest program.

Malahat Nation. No date. Draft letter to Aboriginal Affairs Minister re. Bamberton Town Proposal.

Miller, William. No date. "BC Forests Yield Much Smaller Benefits." Victoria *Times-Colonist.*

Nathan, Holly. 1991. "Natives' Dream Built on Logging." Victoria *Times-Colonist.* 6 October 1991.

—1991. "Natives Seek Share of Forest Industry." Victoria *Times-Colonist.* 6 October 1991.

—1991. "Firms Respond to Winds of Change." Victoria *Times-Colonist.* 6 October 1991.

—1992. "Haida Chief Suspicious of NDP's Priorities." Victoria *Times-Colonist*. 24 April 1992.

—1991. "Deputy Minister Calls for Bigger Native Role in Forestry." Victoria *Times-Colonist*. 18 May 1991.

—1991. "Forests on Reserves 'Mismanaged.'" Victoria *Times-Colonist*. 16 May 1991.

—1992. "Prof Finds Branch of Medicine in Near-lost Native Treatments." Victoria *Times-Colonist*. 11 May 1992.

—1992. "Ditidaht Band to Cash in on Carmanah Tourism with Minimall." Victoria *Times-Colonist*. 23 April 1992.

—1992. "Environmentalists 'Betrayed' by Cashore." Victoria *Times-Colonist*. 10 May 1992.

—1991. "Natives Say Bamberton Idea Infringes on Rights." Victoria *Times-Colonist*. 17 July 1991.

—1992. "Quatsino Band 'Ignored' in Bid to Prevent Logging." Victoria *Times-Colonist*. 19 March 1992.

—1992. "Eco-conference to Hear Positions of Native Leaders." Victoria *Times-Colonist*. 2 May 1992.

—1992. "New NDP Council 'Betrays' Plans for Native Timber Access." Victoria *Times-Colonist*. 20 August 1992.

—1992. "Ucluelet Band's Log-home Scheme Moving Ahead—Other Natives Seek Similar Resource Access." Victoria *Times-Colonist*. April 1992.

—1991. "Give Natives Timber Priority—Report." Victoria *Times-Colonist*. 1 December 1991.

Native Forestry Task Force. 1991. *What We've Heard. Interim Findings of the Task Force on Native Forestry*. Submissions made to task force to April 1, 1991.

Nishga Tribal Council Forestry Committee. No date. *Forests for People: A Nishga Solution*. Pamphlet.

O'Neil, Peter. 1992. "Self-Rule for Natives Arouses Hopes, Doubts." *Vancouver Sun*. 7 October 1992.

Parfitt, Ben. No date. "Value-added Products Vital to Success, Big Forest Firms Told." *Vancouver Sun*.

Schreiner, John. 1992. "Blows Hit BC Forestry Firms." *Financial Post*. 16 June 1992.

Simpson, Scott. 1992. "Anything But . . . Clearcut." *Vancouver Sun*. 25 May 1992.

Smith, Dan. "The Struggle for Self-Rule." *Atkinson Fellowship in Public Policy*, 1991. Copyright: The Atkinson Charitable Foundation. Series in

The Toronto Star. September 1991.

Task Force on Native Forestry. 1991. *Native Forestry in British Columbia— A New Approach.* Final Report submitted to Minister of Forests and Minister of Aboriginal Affairs, November 28, 1991.

Tin-Wis Forestry Conference. 1992. "First Nations, Labor and Environmentalists Building Consensus on the Proposed Forest Stewardship Act." June 1992.

Vancouver Sun. 1992. "Canadians Like Concept, Not Details." (Southam Poll on Native Self-government). 17 February 1992.

Victoria *Times-Colonist.* 1992. "Mulroney 'Redeems Promise' to Natives." 22 September 1992.

Watts, Richard. 1992. "Log Deal 'Creates Overseas Jobs at BC's Expense.'" Victoria *Times-Colonist.* 16 January 1992.

Forest Policy

Rhetoric and Reality

O.R. Travers

Legitimacy requires consent, and consent requires information.
Mikhail Gorbachev

The folly of intelligent people, clear-headed and narrow visioned, has precipitated many catastrophes.
Alfred North Whitehead

IT HAS ALWAYS BEEN MORE FASHIONABLE to talk about British Columbia's dazzling beauty than about its politics, which have little to do with scenery and everything to do with power struggles to share the wealth. When it comes to forest policy, what actually exists is a curious mixture of ideology, legalities and economics that have co-alesced by historical accident rather than as the product of a carefully thought-out set of principles inspired by a wise and farsighted vision-ary. In fact, upon close examination the entire notion of forest stew-ardship in the province, fashionable to many, is based on assumptions of benign paternalism—that the Minister of Forests knows best.

Public policy will determine what the government does in re-sponse to the demands and challenges involved in the future of British Columbia's forests. Ideally, there should be a framework within which decisions are taken and action (or inaction) is pursued by governments to address specific issues or problems. However, what is ideal has seldom been achieved in British Columbia.

BC is a society of recent immigrants who, with the exception of those in the southwest corner of the province, often live in single industry company towns. They have specialized labour requirements, but usually lack stability and tradition. Less than half of all British

Columbians living today in the province were born here; while the remainder have either come from the rest of Canada, or from Great Britain, the United States, Eastern Europe or Asia. As a result, politicians have not been able to rely on constituencies with long traditions when maintaining their power base. Instead, they have to appeal to newcomers who have left somewhere else because they were unable to prosper there. Hence, in practice, the electorate has always valued action over sober reflection.

To make progress in forest policy reform, British Columbians must answer one basic question: What is the role of the provincial government in running the forests? Is the province the land owner? A regulator? An arbitrator? A custodian? A referee in the market place? No one in authority seems to know for sure. Forest bureaucrats often say that people strongly favor continued public ownership of the forests. If this is true, the public should be asking hard questions—including why the forest generates so little government revenue. A concerned public would also insist that the politicians and bureaucrats entrusted to manage, price and sell their forest resources be accountable, through policies and management structures that work for the common good. Instead there is usually silence.

Long-term forestry analyst Pat Marchak, in her chapter for *A History of British Columbia*, personally interviewed two prominent British Columbians, and recorded two very different views on the role of government in forestry. Former Social Credit Minister of Forests, Tom Waterland, in 1978 said:

> The notion of continuous flow forest management is a bad concept. It is really a myth, not used anywhere. The Management can't be done that way. The coast forest is mature, over-mature. We're losing wood if we limit it to that. It is better to cut more and replant . . . There are requirements of processing plants, and the needs of a community to consider. . . . If there's a shortage of supply, somebody will just have to go by the way. It's just a fact of life. There's only so much wood. In order to allocate timber, in order to provide employment, we just have to help the large industries.

While Bob Williams the former New Democratic Party Minister of Forests, said in 1980:

The only genuine case for long term tenures . . . is that the tenure should equal the length of time to amortize the capital investment in plants or the long term debt for major plants. This would rarely exceed twenty years, the same as the average home mortgage. Indeed, the NDP as government found that twelve years was accepted by the industry for new plants in the Babine Forest Products case at Burns Lake, and at the new sawmill in Clinton . . . So the case for these lengthy tenures is very flimsy indeed . . . In fact, such long tenures or extreme privileges in the public forest must be seen to represent the lobbying clout of these semi-monopolists, and the fear of a compliant government to do other than represent the corporate will.

Both views, of course, reflect the policy of each minister's political party. One minister sees the forest as without value until it is cut; for the other the forest represents tremendous wealth which must be prudently used. In the former view the role of government is to rapidly convert the forest into cash and seize the opportunity of present markets, while worrying about tomorrow when it comes. This has been the dominant political view in British Columbia. In the latter view the role of government is to act as a trustee, with the primary obligation being to recover the full value even if some short-term gain has to be sacrificed.

Provincial politicians to date have been content to shape forest policy solely in terms of maximizing their chances of re-election. For many British Columbians, this is no longer good enough, and they are demanding a reformed Forest Act. Since politicians are more inclined to follow than lead, change is going to have to come from the grassroots. Without such support, every noble attempt to do things right will inevitably succumb to the politics of exploitation.

The Making of BC Forest Policy

Development of Ideology and Symbols

Historically, as Robert Nelson shows in his article for *Rethinking the Federal Lands*, public policies have closely followed the broader trends of ideology. Until 1900, the dominant public lands policy in all of North America was quick disposal of lands into private ownership. This earlier policy was supported by classic liberal ideology and assumes a small role for the government. After 1900, the Progressives saw the good society as a highly organized and smoothly working corporate structure. Good individuals were persons who stressed co-operation and self-sacrifice. The effect of their thinking was a new policy that left forest land in government ownership and coincided with the large-scale movement of people from rural to urban life.

The Conservation movement of the early 1900s applied Progressive ideas and beliefs to the field of natural resources. Public ownership of forests was desirable, according to Progressive and Conservationist ideology, for two primary reasons.

First, the private sector was considered too short-sighted and too parochial to rationally manage natural resources. Some conservationists, like Gifford Pinchot, the first Chief Forester of the US Forest Service, "contended that the government would do much better than the private sector in achieving the scientific management of natural resources." Second and more critically, Progressives believed scientific management went hand in hand with large size. Both reasons justified a larger role for government.

Today the Progressive ideology has been replaced by "interest group liberalism." It promotes competition in the political arena as the basic mechanism for allocating resources. On public lands, this ideology is found in the tenets of "multiple use," which seeks to balance competing interests.

Canadians need to be aware of the double bind the government is in concerning the management of forest land. The British Parliamentary system is founded on three distinctive principles of liberal ideology, described by Pat Marchak in *Ideological Perspectives in Canada*.

The first principle is that liberal societies support one or another form of representative government within the framework of the nation state. Secondly, their economies are not directed exclusively or mainly by the governments, and profits from economic transactions may be legally retained. Thirdly, judicial courts evaluate the merits of individual, corporate and government actions with reference to legislation provided by governments, when any two individual or other entities disagree on the nature of their prerogatives.

So an internal contradiction exists at the centre of the liberal view of government, that has the effect of stifling policy development. While governments are the resource managers, and in theory subject to the wishes of the majority, the fact is the economy is not directed or managed by the government. While government can enact legislation to regulate the economy, it cannot restrict the rights and privileges of the major economic institutions. Free enterprise means, essentially, private business or business not directed by governments. In effect, the state ends up politically obligated to protect the interests of the forest industry because of its control on key investment decisions, employment and the general level of economic prosperity. There will not be significant progress in BC forest policy until a way is found to resolve this fundamental difficulty.

Forest policy is based on political symbols that are based on broader liberal ideas and beliefs. For example, forest bureaucrats and politicians once claimed that a sustained yield policy—balancing the cutting rate with the growth rate of timber so there will always be a perpetual timber supply—could, in itself, create and maintain community stability and steady employment. This was Chief Forester C.D. Orchard's vision in the 1940s. His papers show that he had a naïve laissez-faire notion that the "private interest can be made to coincide with public interest and that private interest can be substituted for penalties and coercions."

Moreover, Orchard did not seem to understand the relationship between forest tenure and economic power. He was genuinely surprised by the rush of applicants for Forest Management Licences when the Forest Act was amended in 1947. He wrote:

> Whereas I had thought that, given the authority, we just might induce some public spirited and far sighted operator to take up a forest management license with all its attendant responsibilities, the fact turned to be that almost at once we were deluged with applications. *Industry saw in an assured timber supply a capital gain that I had quite overlooked, and no one in Government or Civil Service detected.* [Author's emphasis].

Politics of Exploitation
The politics of exploitation with provincial sanction have a long history in British Columbia. Law and order between 1850 and 1870 was established on the British Columbia coast with the support of maritime naval power. During the time of James Douglas' governorship,

the British navy was part of the local way of life. Douglas made it clear to the original peoples they were not to ignore his authority or to interfere with settlers. When challenged, Douglas invoked the established Hudson Bay brand of "forest diplomacy" and sent in the gunboats for a show of force. "Forest diplomacy" as exercised by Douglas at Newitty, Kuper Island, Ahousat and Kimsquit, was called "keeping the Indians in awe."

The first commercial enterprise by Europeans was the fur trade. It was finished as a major industry by 1870, but by then it had already set the pattern for resource exploitation, Canada-style. As Peter C. Newman says in *Caesars of the Wilderness*:

> Fur was an extractive industry, carried on by an overseas-based monopoly, strictly for the gain of its private shareholders. That condition—a multinational grabbing Canada's most profitable resources for the one-way benefit of its owners—has characterized Canadian commerce ever since. No nation that has moved past colonial status owns a smaller portion of its profitable assets.

Vancouver Island became a Crown colony in 1849. The mainland of British Columbia became a Crown colony with the capital at New Westminster in 1858. Settlers arrived in large numbers with the authority of the British Empire behind them. Most carried with them a belief of the inherent superiority of their own way of life. Meanwhile, Americans flocked into the province in droves looking for gold, and some stayed, creating the need for a land policy. Thus began the tug of war between idealism and political expediency so common in BC politics. The basic principles of land legislation were established before BC entered Confederation in 1871. Governor Douglas issued a number of Ordinances and Proclamations between 1858 and 1864, including the Gold Mining Act of 1869 which regulated the operation of free miners, and the Land Ordinance of 1870 which provided for pre-emptions (land settled before a survey), the sale of land whether surveyed or unsurveyed, other land leases, and water rights. Building on these, between 1871 and 1913 a land policy was developed in an ad hoc, almost unnoticeable way. Various provincial governments amended or otherwise altered the legislation to meet the needs of the moment. By 1913 there were separate Acts for land disposition, railways, minerals, timber, surveys and water.

To quote Edwin Black in *A History of British Columbia*, there were three constant principles in land disposal legislation: "the wish to encourage settlement, the desire to prevent speculation in public lands, and the acute need for revenue." Therefore, the purpose of Colonial Secretary Edward Bulwer Lytton was fivefold: to reserve to the crown certain rights; to sell land only by public auction; to require prompt cash payment; to survey the land before it was sold; and to survey land before alienation (disposing of it), ensuring its beneficial use.

In the political and economic realities of early British Columbia, it became clear these progressive principles could not be enforced. Settlers quickly discovered they could acquire the land at the upset price set by government because few bidders would show up for the auction. After 1860, Douglas learned that, to encourage the gold miners to settle in BC, he had to forego the cash payments, and he found the survey costs of land before alienation was beyond the colony's ability to pay. He also had to abandon his plan to buy out aboriginal title: instead, he set aside reserves for all lands used or occupied by the First Nations. Increasingly, restrictions had to be eased and resources granted at modest prices. Finally, the lack of money resulted in BC being the only Canadian province to alienate land prior to survey. In other words, people legally acquired title to land, but the map location was unknown by the Victoria Lands Office.

To provide for the transportation needs of settlers and miners, the Smythe provincial government started the practice of subsidizing railway companies. By the 1890s the railway land grants were larger in area than the land taken up by settlers. By 1913, the total land granted to railways was nearly 23 million acres (9.2 million hectares). In the 1871 Terms of Union with Canada, British Columbia gave the Dominion Government a forty-mile wide railway belt through the centre of the province and a two million acre (0.8 million hectare) block on Vancouver Island.

Between 1871 and 1913 timber was disposed of in four ways: by outright sale of timber along with the land; by leasing timber land; by issuing a licence to cut timber; and by selling timber separate from selling the land. Retaining public ownership of the land while selling the timber was the beginning of modern forest policy. By 1913, the government of BC controlled 96% of the forest land in the province.

In 1905, the provincial government moved to take an active role in the exploitation of its forests. The government also acted to arbitrate disputes between different sectors of the industry. To defuse the developing controversy, the government created a Royal Commission to study timber and forestry problems. The Commission was chaired by the government's Minister of Lands, F.J. Fulton. The role of the Commission was to listen to both sides and present the government with a workable consensus. The Commissioners went beyond listening to arguments made by various sectors and attended the 1909 United States National Congress on Natural Resources in Seattle. They later met with Robert H. Campbell, the Dominion Director of Forests in Ottawa, Professor B.E. Fernow, Dean of the University of Toronto School of Forestry, and visited Gifford Pinchot, Chief Forester of the US Forest Service, in Washington DC.

It was the beginning of the long, slow road of official commitment to the ideals of conservation and "regulation" of forest practice. Foresters such as Fernow and Pinchot were articulate early proponents who rationalized forest management on the utilitarian ethic of "the greatest good for the greatest number in the long run." During the 1912 to 1945 period of two world wars and a world-wide economic depression, British Columbia's persuasive timber industries established a government-corporate axis that came to dominate Canadian forestry. While promising sustained yield from government timberlands, politicians lacked the strong commitment to forest conservation that was envisaged when the Forest Service was established in 1912. As Roach and Gillis point out in *A History of British Columbia*, "The exploitive ethic was alive and well in the province, viewing its timber almost solely as a source of revenue and an engine of immediate, unbridled economic growth. Like so many other things started in Canada before 1914, forestry and forest conservation in British Columbia has been transformed from its progressive roots."

Most politicians have won elections in British Columbia on economic development issues. "Improving natural resources" usually has the political meaning of extracting material wealth through hydro power development, logging and mining. Utilization of natural resources has been the great provincial preoccupation ever since and continues as the essence of resource policy in BC. It is a worldview that is materialistic and acquisitive, and it has squandered natural resources as economic power has been extended to the farthest reaches of the province.

The Great Debate over Public versus Private Ownership

The questions about the proper role of the state in forest policy emerged in the debate on public versus private ownership following the Great Depression. In 1939, F.D. Mulholland, a progressive forester of his time, argued for an even division between public and private ownership. He asked, "Who should own the forests? At present the State, that is the province, owns 93 percent of the forest land of British Columbia. In forest ownership we are comparable with Soviet Russia. If you lean towards communism you, no doubt, regard this condition with satisfaction. It is nationalization of natural resources, if we can regard the Province as a sovereign State for this purpose."

In 1946, Chief Forester Orchard outlined four options for the role of the government in forest management.

1. The State should be interested only in such forest management as is related to parks, watersheds and sub-profit lands. It should own no productive forest land.
2. The State should own and manage forests on a profit basis, but not necessarily to the exclusion of private interests.
3. The State should own lands and grow timber but hold itself severely aloof from any form of logging and manufacturing and from any regulation of private industry.
4. The State should own and manage forest lands for the exclusive benefit of private interest, the product to be allotted to operators and motivate owners under a system of regulated cutting.

Orchard had an economic blind spot which has plagued forest policy ever since. He appeared to believe that forest land had no value and there was no alternative but to appeal to altruism. It is as though government had a solemn duty to come up with an institutional arrangement to persuade forest companies to take this useless land which only had value when timber was cut. On the four options noted above, Orchard said:

Underlying most of these shades of opinion appears to be a deep concern for immediate profit and a failure to realize that forest management is never practiced for a day. The virgin forest, or the well managed forest, will always have in it more than enough timber for the immediate present. The same may be said even of the present day abused forests of Canada. Management is for coming generations and therein is probably the most completely altruistic undertaking of either State or private endeavor.

In 1947, Mulholland in a rejoinder continued his arguments:

In the past years there has been a good deal of propaganda designed to create the impression that without more Government ownership and control of forest lands, we are headed for a catastrophe. That is I think largely a reflection of similar propaganda in the United States. There it is understandable, with 80 percent in State ownership. In British Columbia to try to increase the interest of the private citizen in raising the crop is so natural to the Province, and so important for its prosperity.

Something of a compromise stance was reached with the 1947 session of the legislature, which forever changed forest policy in British Columbia. R.C. Telford of the Forest Service said the intent of the Forest Act amendments was to carry on "the long established principle of government ownership and final responsibility for forest lands, but by providing for 'Forest Management Licences' with long term tenure and relief from unreasonable carrying charges, make possible industry interest in the production of forest crops."

The debate continued. The 1956 Royal Commission on Forest Resources was created to assess the policy of awarding Tree Farm Licences. At the center of the controversy were two prominent men: H.R. MacMillan, founder of MacMillan Bloedel, and C.D. Orchard. As *Three Men and a Forester* recorded, "Ironically, Orchard the public servant and chief forester spoke for the selfish interests of big industry, while MacMillan, the arch-industrialist, was a staunch defender of small business and a protector of public property."

MacMillan provided what turned out to be very prophetic testimony at the Royal Commission:

It will be a sorry day . . . (for) British Columbians when the forest industry here consists chiefly of a few very big companies, holding most of the good timber—or pretty near all of it—and good growing sites to the disadvantage of the most hard working, virile, versatile, and ingenious element of our population, the independent logger and the small mill man.

A few companies would acquire control of resources and form a monopoly. It will be managed by professional bureaucrats, fixers with a penthouse viewpoint who, never having had rain in their lunch buckets, would abuse the forest. Public interest would be victimized because the citizen business needed to provide the efficiency of competition would be denied logs and thereby be prevented from penetration of the market.

Chief Justice Sloan ignored MacMillan's testimony. His predictions have become true and pervade all aspects of forestry today. He foresaw the inequities and injustices and the associated poor forest practices that exist forty years later. The sad reality is the longer we postpone the change, the further we will fall, and the more expensive the recovery will be. How long must we wait to start the long hard road back?

The Effect of Political Myths and Symbols

One of the major difficulties ordinary British Columbians have in clearly grasping forestry issues is the pervasive use of political symbols and the propensity of politicians to grossly exaggerate the benefits from forest cutting. To understand the importance of political symbols in manipulating public opinion it helps to view human behaviour through the eyes of a political scientist such as Murray Edelman:

Political history is largely an account of mass violence and the expenditures of vast resources to cope with mythical fears and hopes. At the same time, large groups of people remain quiescent under noxiously oppressive conditions and sometimes passionately defend the very social conditions that deprive or degrade them.

Systematic research in political science in recent decades has repeatedly called attention to the wide gulf that exists between our solemnly taught common sense assumptions about what political institutions should do and what they actually do. The success of using political symbols and myths in British Columbia forest policy to create a sense of false security is explained by Jeremy Wilson in *Forest Conservation in British Columbia, 1935-1985*:

> ... the powerful positive symbolism associated with sustained yield contributed to the drift into complacency ... An aroused public is placated by a policy response that is at least as strong on symbolism as on substance. In the mood of quiescence that follows, authorities proceed, largely unnoticed by the public to reverse or neutralize the putative intent of the policy ... It argues that in large part because of these symbols, that policy became a kind of security blanket for British Colombians of the 1950s and 1960s. During these years it seems to have been widely believed that sustained yield guaranteed that the province's forests would produce a perpetual even-flow (or perhaps a perpetually increasing flow) of timber wealth.

This is exactly the kind of effect described by Murray Edelman:

> Language forms and terms reinforce the reassuring perspectives established through other political symbols, subtly interweaving with action to help shape values, norms, and assumptions about future possibilities ... Trite phrases may be used as incantations, serving to dull the critical faculties.

Fifty Cents of Every Dollar

One of the best examples of an influential "incantation" is the one about the forest industry's contribution to the BC economy. In the July-August 1990 issue of *Forest Planning Canada*, editor Bob Nixon announced, "Death of Another Forestry Myth: Fifty Cents of Every Dollar":

> A myth created long ago by the forest industry of BC claims that "50 cents of every dollar comes from the forest industry," repeated so often it was perceived as truth. But now even the Council of Forest Industries of BC recognizes the futility of attempting to maintain this particular illusion any longer.

Its death came on May 12, 1990 in Squamish during a conference entitled "Future of Forestry," sponsored by the Sea to Sky Economic Development Commission, and the industry's own hand, Dick Bryan, Manager, Economics, Statistics and Energy of the Council of Forest Industries, said "You've probably heard the old saying about 50 cents out of every dollar generated in BC coming from the forest industry. Well, that's not true today and actually, I'm not sure it ever was . . ."

Even though it produces nearly half the products manufactured in the province, in total, forest sector activity accounts for about 25 percent of the gross domestic product. So if you want to express the industry's economic significance in terms of cents out of every dollar you're looking at about 25 cents out of every dollar.

Even this 100 percent downward revision has turned out to be an exaggeration. The 1991 BC Forest Resources Commission stated that, "In 1989, the forest sector's total (direct, indirect, and induced) contribution to the provincial economy was 16.9 per cent. On the same basis it contributed 14.4 percent of total employment." So it turns out the original claim of economic importance was a 300 percent exaggeration. This revelation has been hard to accept for many in the political culture of BC, where the exploitation of rich natural resources has been considered an inherent right.

Competitive Position
Another favourite forest industry and government political symbol is "our competitive position." It is reinforced with enough conviction to persuade people that we actually take the issue seriously and is usually accompanied by dire warnings intended to put fear in the hearts of those who dare to speak out.

However, being competitive in world markets is not only a question of having an abundant supply of high quality natural resources at artificially low prices. In fact, nations like Japan and Germany with very few natural resources of their own are amongst the most competitive nations in the world. Recent information shows that Canada's competitive position, largely resulting from an abundance of high quality natural resources that Canadians did very little to create, is slipping badly.

In 1990, the Swiss-based World Economic Forum ranked Canada

fifth in international competitiveness in their annual report. A year earlier Canada was fourth. By 1992, Canada had dropped to eleventh, but Japan and Germany had increased to first and second. The factors assessed in this ranking are a country's domestic economic strength, internationalization, government, infrastructure, finance, management, science and technology, and work force. In 1990, Canada got good grades for political stability, standard of living, banking system, natural resources and people. However, Canada got low marks in areas that will be increasingly important in the future, such as product quality, innovation, research and development, the entrepreneurial drive of businesspeople and the willingness to modify methods for foreign markets. In 1990, out of thirty-seven countries ranked, Canada was twentieth for restructuring of the economy to adapt for long-term competitiveness, twenty-first for strength of manufacturing base, nineteenth for international experience and senior management, twelfth for quality of education system and fourteenth for work attitudes of young people. The message is clear. Canada's competitive position is slipping rapidly, and increasing the rate of logging in the province will not in itself solve anything. In fact, it will make our vulnerability even worse. The de-industrialization of Canada has already begun.

Reality Check: The Facts of Forestry

People who question forest policy must be both skillful and persistent as they probe a system sufficiently complex and uncertain to overwhelm most potential critics. This search is further complicated by economic and technological assumptions accepted in earlier times that are no longer valid. Continuing change also makes it difficult to predict what will take their place. Moreover, the enormous discretionary policy-making power of the Minister, usually exercised behind closed doors, means the rationale for decisions, if there is one, may never be known.

To make good public policy requires, in Sir Geoffrey Vickers' words, the "setting of governing relations or norms." It is not merely the setting of goals, objectives or ends. The challenge is to know where you are, and to know where you are going—and you can start the process by making a reality check of the current state of affairs. This is not easy in BC, given the abundant misinformation. To help out, data has been assembled from a large number of official government reports showing the actual social and economic benefits of our forest policy. This is where we are.

Size and Control of BC's Forests

While BC is celebrated for its forests, it is not generally known that only half the province is covered by trees. This is important when it comes to assessing the extent of our options. Data from the Ministry of Forests (Table 1; Figures 1 and 2) illustrates this point and is interesting because of what it shows about potential control.

Table 1

HOW MUCH FOREST LAND THERE IS AND WHO CONTROLS IT

	Forest Land (hectares)	Non-forest Land (hectares)	Total (hectares)	Percent of BC %
Public Forest Land				
Timber Supply Areas	39.5	35.5	75.0	79.0
Tree Farm Licenses	3.7	3.2	6.9	7.3
Provincial/Federal Park	2.0	3.8	5.8	6.1
Private Forest Land	2.7	0.0	2.7	2.8
Subtotal Forest	47.9	42.5	90.4	95.2 %
Agriculture, Urban etc.	0.0	4.2	4.2	4.4
Total Land	47.9	46.7	94.6 million	
Percentage of BC	50.6%	49.4%		100.0%

Source: Ministry of Forests, Park Data Book

Clearly, the most valuable forest land is concentrated in private ownership and Tree Farm Licenses (TFLs). Remembering that only 50.6% of BC is forest land throws a new light on industrial proposals to allocate forest land for commercial use. For example, Mike Apsey, President and Chief Executive Officer of the Council of Forest Industries has said:

Figure 1

FOREST AND NON-FOREST LAND AND WHERE IT IS IN BC, 1989

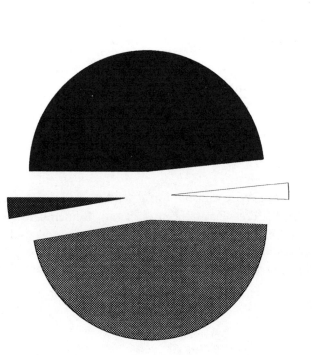

■ Forests on Managed Public and Private Land

□ Forests on Provincial and Federal Park Land

▨ Non Forest Crown Land

■ Other Land (Urban, Agriculture etc.)

Source: Ministry of Forests, Provincial Parks Data Book

Figure 2

OWNERSHIP AND CONTROL OF ALL LAND IN BC, 1984

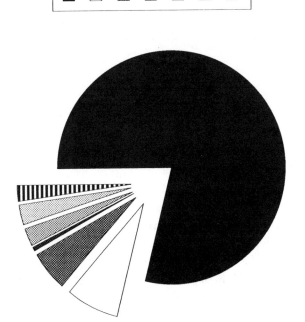

Legend:
- ■ Timber Supply Areas
- ☐ Tree Farm Licences
- ▦ Provincial Parks
- ■ Federal Parks
- ▨ Other Crown
- ░ Private Forests
- ☰ Urban and Agriculture

Source: Ministry of Forests, Ministry of Crown Lands

We envisage a defined commercial forest land base of about 30 percent of the land in the province (British Columbia) under public and private ownership, supporting a world competitive and environmentally attuned industry.

This proposal appears to be very modest, especially in that only 24 percent of the area of the province is presently included in allowable annual cut calculations. But the real effect of classifying 30 percent of the province as commercial forest land would be to put under forest industry control all the valuable forest land in the province, probably under the control of timber monopolies. This land is also the best land for most other uses. As well, COFI seems to want to control all the private and commercial forest land—whether good site, medium site or operable poor site land (see Table 2).

Table 2

QUALITY OF FOREST LAND AND COFI OBJECTIVES

Quality Class TSA and TFL	% of BC	COFI Objective
Good Site	3.5%	3.5%
Medium Site	16.5%	16.5%
Poor Site	21.0%	7.2%
Low Site	3.3%	0.0%
Other Land		
Private	2.8%	2.8%
Park	2.2%	0.0%
Total	49.3%	30.0%

Source: 1984 Forest and Range Analysis, Ministry of Forests

Accelerating Rate of Logging

The history of forest policy in British Columbia can be characterized as a "pull and tug" between the ideal and the real. Every effort to establish the principles of forest conservation sooner or later succumbed to the politics of exploitation. In 1912 progressive principles appeared to be firmly established when the BC Forest Service was established. Thomas R. Roach writes in *Stewards of the People's Wealth*:

The provincial government had checked the transfer of public forest lands to private companies in 1905, leaving BC with one of the highest percentages of commercial forest under government control in the world . . . Relative newcomers to the North American forest conservation movement, British Columbia foresters were not only able to use the experience of others to justify radical action to curb industrial excesses, but they were able to avoid the pitfalls of developing a forestry system in a region traditionally controlled by private interests operating on public lands . . . The history of forestry in the province was in many ways a capstone to the early twentieth-century conservation movement.

And progressive thinking seemed reaffirmed in 1956 when the chairman of the Royal Commission, Chief Justice Sloan wrote: "It is in my opinion, a negation of the principle of sustained yield, and the social aspects implicit therein, to manage a forest intentionally in order to harvest large areas of it at infrequent intervals, even up to a whole rotation, with long periods intervening without any production."

The intent was to regulate the cut of public timber to provide a steady supply of wood, but to leave the private timber outside of forest management units unregulated. This is where one enters an Orwellian world and words have their opposite meaning (see Table 3, Figures 3 and 4).

Relative to their area, there has been a very disproportionate supply of timber from private forests. Since 1912 these lands have contributed 1.0 billion cubic metres of the total 2.7 billion cut in the province. This is the formally "unregulated" cut. However, this unregulated cutting of private timber has remained a relatively stable 12 to 15 million cubic metres a year. At the same time the "regulated" cut in sustained yield units has increased from 3 million cubic metres a year in 1945 to nearly 80 million cubic metres in 1988. Fifty percent of all the public timber cut has been felled in the last 13 years. The most rapid acceleration has primarily been in the BC interior, with 50 percent of the total cut being done since 1977—the northern regions of Prince George and Prince Rupert have an even faster acceleration rate.

Table 3

AMOUNT AND TIMING OF TIMBER CUT IN BC 1911 TO 1991

Ownership Categories	50% of Total Cut	Total Volume Cut
Private*	Since 1960	1.0 billion m³
Crown**	Since 1978	1.7 billion m³
		2.7 billion m³
Coast (All Tenures)	Since 1967	1.4 billion m³
Interior (All Tenures)	Since 1977	1.3 billion m³
		2. 7 billion m³

Crown Grant, Timber Licences (Old Temporary Tenures)
** *Timber Supply Areas and Tree Farm Licences*

Much of this acceleration has been above the Ministry of Forests' own estimates of the sustainable yield. The long run sustainable cut, using questionable Forest Service assumptions, indicates a cut of about 60 million cubic metres a year. However, this has not prevented the volume allocated in forest tenures (the committed cut) increasing about 1 million cubic metres a year since 1976 (see Table 4).

Table 4

TOTAL COMMITTED YEARLY CUT OF PUBLIC FORESTS

Year	Committed AAC (million m³)
1975	60.8
1980	66.7
1985	67.3
1990	74.3
1991	74.5

Source: Ministry of Forests Annual Reports

Method of Cutting: Clearcut versus Selection

Records of the yearly area logged have been kept in Forest Service Annual Reports since 1970. These show the accelerating speed of

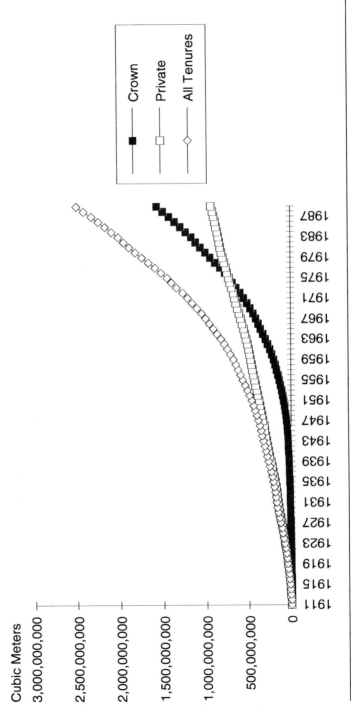

Figure 3

CUMULATIVE AMOUNT OF TIMBER CUT IN CROWN AND PRIVATE FORESTS, 1911–1989

Cubic Meters

3,000,000,000

2,500,000,000

2,000,000,000

1,500,000,000

1,000,000,000

500,000,000

0

1911 1915 1919 1923 1927 1931 1935 1939 1943 1947 1951 1955 1959 1963 1967 1971 1975 1979 1983 1987

Crown

Private

All Tenures

Source: Ministry of Forests Annual Reports

Figure 4

CUMULATIVE AMOUNT OF TIMBER CUT IN COASTAL AND INTERIOR FORESTS, 1911–1989

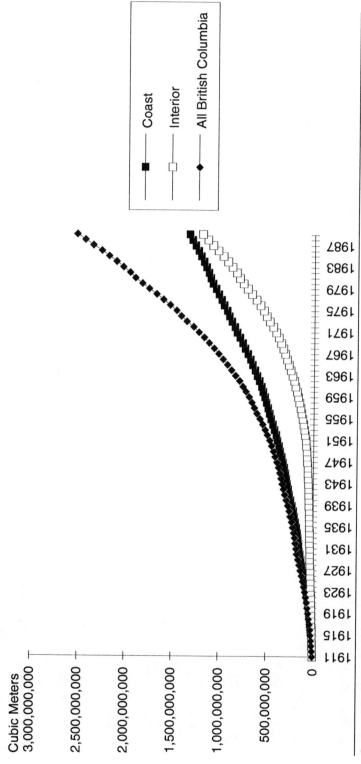

Source: Ministry of Forests Annual Reports

logging is paralleled by an increased use of clearcutting rather than use of the selection logging method (see Table 5, and Figure 5).

Table 5

AMOUNTS OF CLEARCUT VS. SELECTION LOGGING IN BC, 1971–1991

Year	Clearcut	Selection	Total Hectares
1971	135,000	30,000	165,000
1975	100,000	19,000	119,000
1980	187,000	29,000	216,000
1990	199,000	19,000	218,000
1991	166,000	15,000	181,000
Sum 1971-91	3,448,176	518,497	3,966,673 Hectares
Average	172,000	26,000	198,000 Hectares

Source: Ministry of Forests Annual Reports

The year 1988-89 holds the record for the highest volume of timber cut with 90 million cubic metres. In this year, there were 271,000 hectares logged (247,000 clearcut, 24,000 selection). To put it another way, 1055 square miles were logged in 1988-89. The area logged using selection methods continues to decline. In 1990-91, only 15,000 hectares were felled using this method— an all time low since records began.

Undervaluation of Public Timber

Perhaps the major policy issue concerns the valuation of public timber. More than any other issue, the way this question is answered will determine the value of benefits produced. Ideally, fair market value for stumpage should be the standard. Stumpage is the price (like a "tax") logging companies pay to the government for the right to cut public timber. While fairly straightforward in concept, achieving fair market value for stumpage means meeting numerous conditions, which is frustrated by the market control created by the regional timber monopolies and a characteristic of the existing forest tenure system. The fact is that open competitive markets for logs have been the exception in BC. By 1980, the independent logging sector had all but diminished or disappeared, though it has since been partially revived to 10 percent

Figure 5

PROPORTION OF AREA SELECTION LOGGED AND CLEARCUT COMPARED TO VOLUME CUT IN BC, 1911–1989

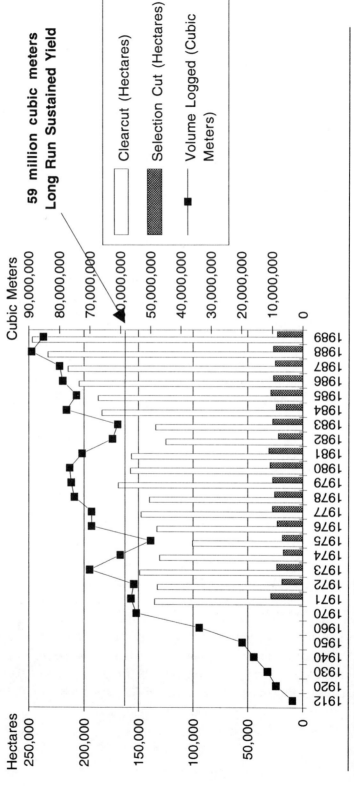

of the annual volume logged by the introduction of the Small Business Forest Enterprise Program.

The under-valuation of logs sold from public land is an issue of urgent public concern. Ian Mahood in *Three Men and a Forester* says:

> There are many sad and outrageous aspects to the situation that has developed, particularly since 1975. One is the plundering of the public purse. It is difficult to calculate how much public equity has been lost by under-assessed sales of public timber, under-measure of public property and other manipulations of the fixers. If one compares stumpage and prices collected on public timber sales in BC and others on a global market, a $1 billion dollar reduction in public revenue seems reasonable. Some of this shortfall has been collected during the last couple of years, thanks to the efforts of the US lumber producers who threatened to collect a tariff on our lumber if we didn't come to our senses and begin collecting it here.

It is clear that where logs are allocated using public auction, dramatic increases in value are possible. It is perplexing, to say the least, in these times of seven-figure provincial government deficits, that little political interest is shown in moving towards a log market, even by so-called free enterprise politicians. Economists generally state that about 50 percent of the annual volume logged must be sold in log markets before true values are generated. Even the Small Business Forest Enterprise Program, representing 10 percent of the volume cut, produces values two to three times the administered prices for stumpage established by Forest Service bureaucrats.

The Economic Performance of the Forest Industry

The forest products economy in BC produces high volumes of a very narrow range of forest products, and is very sensitive to economic cycles. In 1990-91, after some very good years, the industry reported an overall loss of an estimated $900 million. While the gross value in forest products has been very large—over $10 billion annually in recent years—a careful analysis shows that BC is among the lowest producers of value per cubic metre of any forest economy in the world (see Table 6).

Table 6

VALUE AND JOBS FOR TIMBER CUT IN BC COMPARED TO OTHER COUNTRIES, 1984

Country	Volume logged Million m³	Value Added/m³ Through Processing	Jobs/1000 m³
BC	74.6	$ 56.21	1.05
Other Canada	86.3	$110.57	2.20
Canada	160.9	$ 83.39	1.62
United States	410.0	$173.81	3.55
New Zealand	5.3	$170.88	5.00
Sweden	56.0	$ 79.49	2.52
Switzerland	7.0	N/A	11.41

Sources: Statistics Canada 25-201, 25-202; USDA, Forest Service PA-1384; New Zealand: New Zealand Forest Service, 1987; Svensk Skog, National Board of Forestry, 1987; Swiss Federal Office of Forestry, FAO Forestry Paper 86

Table 7 shows that BC compares poorly even to Canadian provinces with nowhere near the same quality and volume of timber:

Table 7

VALUE AND JOBS FOR TIMBER CUT IN BC COMPARED TO OTHER CANADIAN PROVINCES, 1986

Province	Volume logged Thousand m³	Value Added/m³ Through Processing	Jobs/1000 m³
Newfoundland	1966	N/A	N/A
PEI	160	62.50	N/A
Nova Scotia	3533	102.75	2.00
New Brunswick	8380	87.28	1.73
Quebec	32526	123.84	2.51
Ontario	27859	138.95	2.86
Manitoba	1590	128.81	2.94
Saskatchewan	3339	N/A	N/A
Alberta	10298	54.50	1.01
BC	77503	61.19	1.04
Canada	167154	90.52	1.72

Source: Selected Forestry Statistics E-X-44

The conclusion is clear. In British Columbia the present structure of the forest industry generates very little value and very few jobs for each unit of wood processed. If we truly want to employ more people, we must restructure the industry to add more value. As Figure 6 demonstrates, provinces like Quebec and Ontario capture more value and create more employment because they have a more diversified solid forest products industry, especially in the paper and allied sector. In comparison many BC pulp mills, especially in the Interior, do not even have paper machines.

The Forest Industry's Contribution to BC's Economy

Since 1960 the volume of timber cut has increased from about 30 million cubic metres to 90 million cubic metres. During the same period, the direct forest industry percent of the Gross Domestic Product declined from a high of 15 percent down to about 10 percent, with a low of 8 percent in the 1982 recession. In other words, the value of the BC economy in general has been increasing faster than the contribution of the forest industry despite the huge increases in annual volume logged. Figure 7 shows that the contribution of the logging component and the pulp and paper component to the economy has been very flat, while the solid wood component that produces lumber and other solid wood products (GDP Wood Industries) grew very rapidly, with a low in the 1982 recession, and a peak in 1988 when the Canadian dollar had the lowest value relative to the US dollar.

Government Revenue from the Forest Products Industry

Government revenue has its peaks and valleys consistent with the highs and lows of economic cycles. The largest direct contribution of the forest industry to provincial revenue is for stumpage payable to the Ministry of Forests for the net value of the trees it cuts. However when logs are not sold by public auction and therefore do not reflect true market value, the actual return to the public treasury can be very small. In fact, most regions of the province in recent years have not generated enough revenue to allow the Ministry of Forests to pay its management and administration costs (see Table 8).

Figure 6

VALUE FOR TIMBER CUT AND MILLED COMPARED BY PROVINCE, 1986

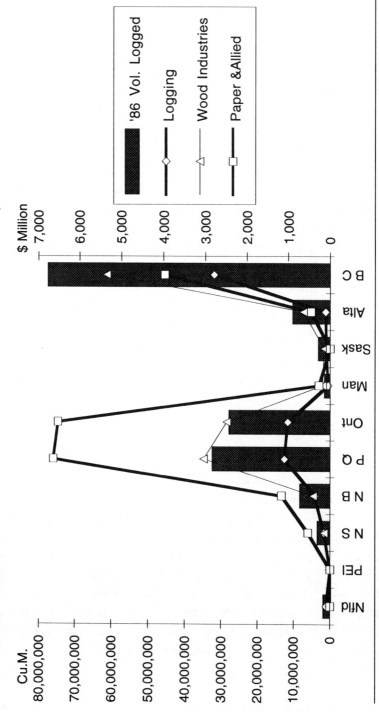

Source: Selected Forestry Statistics, E-X-40

Figure 7

COMPARISON OF THE AMOUNT LOGGED AGAINST THE GROSS DOMESTIC PRODUCT IN LOGGING, WOOD INDUSTRIES AND PAPER AND ALLIED PRODUCTS, 1961–1989 (in 1981 dollars)

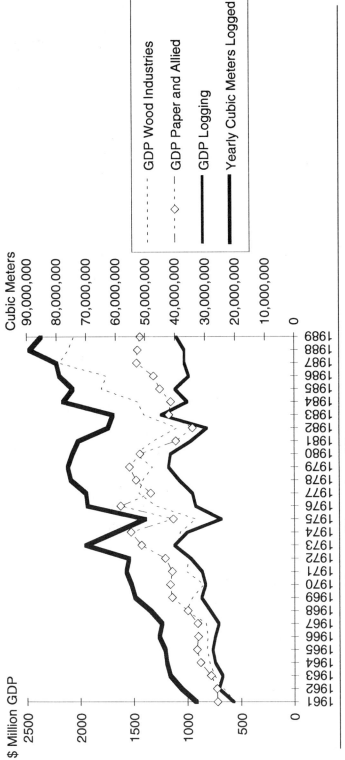

Source: BC Economic Accounts, Ministry of Forests Annual Reports

Table 8

NET REVENUE FROM PUBLICLY OWNED TIMBER IN BC, 1990–91

Region	Amount Cut Crown Land	Net Revenue*	$ Income per m³**
Cariboo	7.7	5.9 loss	0.77 loss
Kamloops	7.4	27.7 loss	3.75 loss
Nelson	5.4	38.1 loss	7.00 loss
Prince George	15.9	16.7	1.05 profit
Prince Rupert	9.3	10.8 loss	1.16 loss
Vancouver	21.2	76.2	3.60 profit
1990-91 overall profit	66.9 million m³	$10.4 million	$0.13 per m³

* Includes Victoria Overhead   ** Figures averaged
Source: Ministry of Forests Annual Reports

These marginal returns are not unusual. In fact, between 1982 and 1987 the Ministry of Forests lost $1.1 billion. The loss ended in 1987 when the US government agreed in the Memorandum of Understanding on Softwood Lumber that BC could collect the additional revenue in lieu of a tariff being imposed to offset undervalued stumpage. Accordingly, as Table 9 shows, after 1987 gross stumpage revenues increased from about $200 million to $600 million annually.

Of course there are also indirect returns to the public treasury from other sources of revenue. Figure 8 illustrates that the share of provincial revenue contributed by the timber sector from all direct and indirect sources (personal income tax, sales tax etc., including stumpage revenue) to all levels of government in BC has ranged between about seven percent and sixteen percent. It is interesting to note that the contribution of the forest industry to government revenue is in about the same proportion as its share of the provincial GDP.

It is popular for the Ministry of Forests (MOF) and the forest corporations to overstate their case by implying that the taxes and other forest revenues they pay are solely responsible for paying the costs of social services. For instance, the District Manager for Williams Lake wrote in a 1992 statement in the *Big Creek Bugle*:

Table 9

COMPARISON OF AMOUNT LOGGED ON CROWN LAND TO NET RETURN TO THE PROVINCIAL TREASURY, 1970–91

Year	Amount Logged Crown Land million m³	Total Net Revenue $ million	Direct Net Revenue per m³
1970	46.1 million	20.7 million	$0.45
1975	43.6	68.0	$1.56
1980	68.6	349.5	$5.10
1981	65.4	94.1	$1.44
1982	57.6	188.8 loss	$3.28 loss
1983	56.8	217.9 loss	$3.84 loss
1984	71.2	137.1 loss	$1.92 loss
1985	68.6	137.3 loss	$2.00 loss
1986	72.3	234.0 loss	$3.23 loss
1987	73.3	192.1 loss	$2.62 loss
1988	80.8	1.4	$ 0.02
1989	75.6	142.1	$1.88
1990	77.6	113.2	$1.46
1991	66.9	10.4	$0.13

Source: Ministry of Forests Annual Reports

The MOF acknowledges the importance of the forest industry to the Province's economy and employment. Job losses due to automation, the alienation of forest land, or for any other reason concerns the MOF, just as it should concern all British Columbians. A weakening of the forest industry has the potential to adversely affect the Provincial economy and our standard of living, including education, health care, social services, etc.

This argument assumes, first, that the timber sector always produces a surplus of government revenue above the costs of management and administration, and second, that non-timber uses for the same forest would not. Both assumptions need to be examined. Figure 9 shows that total revenue of the Forest Service has never exceeded 10

Figure 8

FOREST SERVICE REVENUE AS PERCENT OF BC GOVERNMENT REVENUE, COMPARED TO TOTAL FOREST REVENUE AS PERCENT OF ALL FEDERAL, PROVINCIAL AND MUNICIPAL REVENUE IN BC, 1976–1988

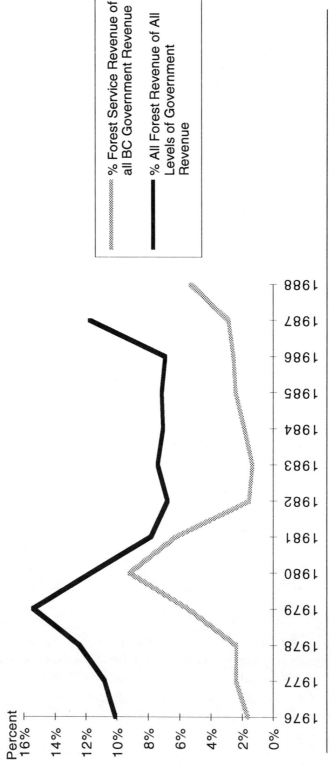

% Forest Service Revenue of all BC Government Revenue

% All Forest Revenue of All Levels of Government Revenue

Source: Price Waterhouse, Ministry of Forests Annual Reports

Figure 9

FOREST SERVICE REVENUE AS A PERCENT OF TOTAL BC GOVERNMENT REVENUE, COMPARED TO PERCENT NET FOREST SERVICE REVENUE AVAILABLE FOR SPENDING ON HEALTH, EDUCATION AND SOCIAL SERVICES, 1950–1989

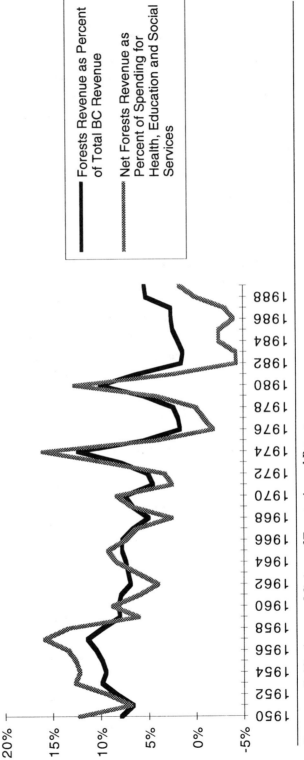

Forests Revenue as Percent of Total BC Revenue

Net Forests Revenue as Percent of Spending for Health, Education and Social Services

Source: BC Public Accounts, Ministry of Forests Annual Reports

percent of the gross provincial revenue, and that net Forest Service revenue—the amount available for spending on other government services—has never exceeded 15 percent of the total health, education and social services expenditures. Moreover, since 1976, Forest Service expenditures *have exceeded* revenue in 8 years out of the last 14. In other words, the Forest Service has not had enough revenue to pay its own bills, and did not generate a surplus to help pay for social services provided by other government departments.

Forest Industry Jobs
Employment is one of the main reasons politicians frequently give for the cutting and processing of timber. This is often used to justify the need for logging in land use debates. Until recently, the number of jobs have increased with the rate of cut. In 1944, 15 million cubic metres were logged, and there were 30,000 direct forest industry employees. In 1991, 85,000 million cubic metres were logged, and there were 85,000 direct employees. However, automation has reduced the number of jobs per thousand cubic metres logged from two to one, with further reductions anticipated. (See Figure 10 and 11). In March 1992, Bob Elton of Price Waterhouse predicted:

> . . . before the last recession there were 95,000 people employed in this industry and during the 1981 to 1982 period, 20,000 jobs were lost; 20 percent of the total. It is true that 5,000 or so jobs were regained through recovery but clearly there was a large net loss. In 1990, 81,000 people were directly employed in the industry compared to 95,000 at the start of the last recession.
>
> It is not possible to predict with certainty but we expect that the recession we are undergoing now will ultimately result in a drop in employment which is at least as large as that experienced in 1981/82. In other words, we expect about 15,000 to 20,000 job losses. This may even prove to be conservative. This type of restructuring is not avoidable, and indeed if the industry tries to avoid or even delay it, the ultimate job losses may even be much greater.

Figure 10

AMOUNT LOGGED IN BC COMPARED TO DIRECT FOREST INDUSTRY JOBS, 1944–1989

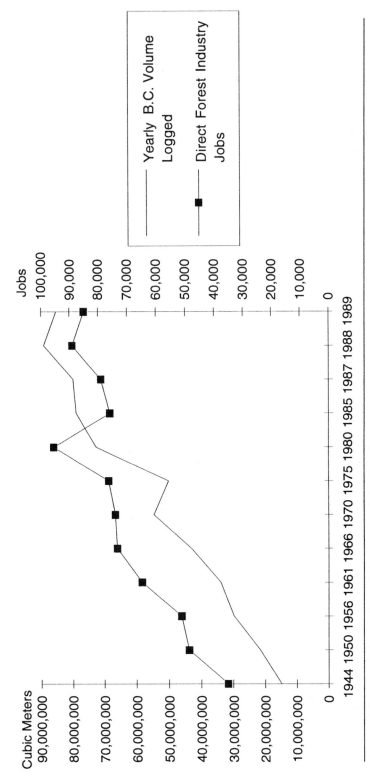

Sources: *Ministry of Forests Annual Reports, Statistics Canada*

Figure 11

AMOUNT LOGGED COMPARED TO DIRECT FOREST INDUSTRY JOBS PER THOUSAND CUBIC METRES IN BC, 1961–1989

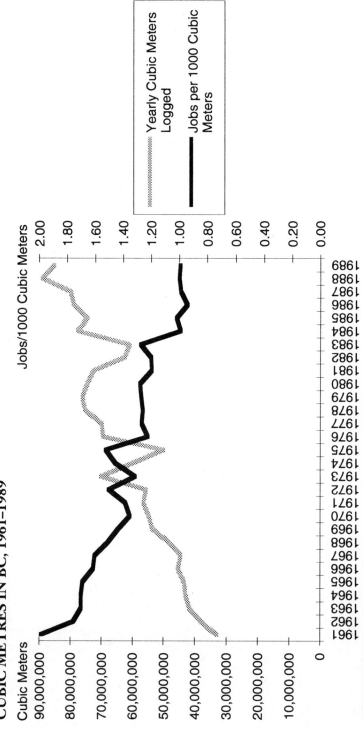

Sources: Ministry of Forests Annual Reports, Statistics Canada

The Forest Industry's Effect on Community Stability

There are two economies in British Columbia. More than half of all British Columbians live in the southwest corner of the province with a diversified economy that has a modern transportation and communications infrastructure, convenient access and proximity to other population centers, developed business support services and a large local market. This is where all BC's population growth—about 10 percent—occurred in the 1980s. Meanwhile, in the other economy of the province, the population actually declined; namely in the northeast, central Interior, southern Interior, north Coast, central Coast, and the Kootenay regions of the province (See Figure 12).

The lack of diversification in rural economies is a major concern. The BC Central Credit Union stated in May 1990:

> About half of all British Columbians live in the Lower Mainland. For these people, improvements in both economic growth and stability have been possible because of the successful diversification of their economy. For them, prospects are for even greater gains ahead.

But the news is less positive for the 1.5 million British Columbians who live outside the urban southwest. In the regions, diversification has played a much smaller role in economic development. A continued reliance on traditional resource industries is hampering both current and future economic growth and stability.

Economies in single industry towns are fragile. In a 1987 report by the Canada Employment and Immigration Advisory Council titled *Canada's Single-Industry Communities: A Proud Determination to Survive*, the authors state:

> Plant closures are not new in Canada. The remnants of more than 400 ghost towns across the country stand as mute reminders of communities that became the victims of exhausted ore bodies, denuded forests, declining fish stocks, changes in transportation patterns, and other adversities. The difference today is that residents of single industry towns do not view the demise of their communities as inevitable. They feel that with hard work and more planning, their towns can be saved. And even if abandonment is the only alternative, they believe the trauma involved can be reduced by preparing for it in advance.

Figure 12

POPULATION GAINS AND LOSSES BY REGION IN BC, 1981–1989

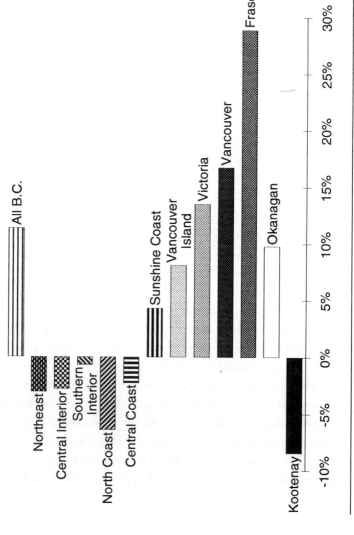

Source: Central Statistics Bureau

Given that single industry towns are vulnerable to boom and bust, it is difficult to understand why forest industry advocates make such a virtue out of them, unless it is solely to protect their short-term interests. For instance, the Forest Resources Commission provided a rationale for even greater overcutting of timber for what it says are 200 forest dependent communities (the above federal advisory council says there are 26 forest industry dependent communities with greater than 1000 people). The Commission said in an open letter:

> Because of the overwhelming employment dependency on the forest industry in so many of these communities, great care must be taken when considering actions that may result in a significant reduction in forest industry activity. Failure to do so could result in major social dislocation and community upheaval outside of the Lower Mainland. In fact, based on the evidence contained in this report, we should be actively considering policies aimed at maintaining or even increasing timber harvest levels . . .

This rationale is poorly thought out and shortsighted. White, Duke and Fong in a 1989 Forestry Canada Information Report, *The Influence of Forest Sector Dependence on the Socio-Economic Characteristics of Rural British Columbia*, stated it much better:

> Outside of the major southwestern urban centers, 30 percent of all communities and regions are dependent solely on the forest sector for their economic well being, and another 40 percent rely on the forest industry as one of their top three employers. During times of economic growth, this dependence has been a boon for rural British Columbia as the forest industry has consistently paid high wages and salaries. However, this dependence has left much of the province vulnerable to economic downturns, such as the recession of the early 1980s.

Their study defined economic dependence for incorporated communities and territorial subdivisions outside of southwest BC as follows:

— **Forest-specialized:** The forest sector alone dominates the economic base.

— **Dual:** The forest and one other sector dominate the economic base.

— **Diversified:** At least three sectors dominate the economic base and the forest sector is one of these.

— **Specialized non-forest:** One non-forest sector dominates the economic base.

— **Minor or no forest:** There is no forest sector component in the top five stated occupations.

They were able to show that single industry dominance was not beneficial for either employment or for population stability. By analyzing communities in 1981, as the economy headed into the recession, and in 1986, coming out of it, they found that diversification buffers a community or region when one of the leading sectors falters. You can see by looking at Tables 10 and 11 that as economic diversification increased, the employment multipliers increased. One option for diversification is to add other basic industries; another is to establish services within the community.

The ratio of incoming people to outgoing people is a measure of community stability. A ratio of greater than one indicates that more people are entering (inmigrants) than are leaving (outmigrants). A ratio of less than one indicates the reverse. During the 1980s, more people in rural BC left rather than entered resource communities.

Table 10

1981 EMPLOYMENT MULTIPLIERS AND RATIO OF INMIGRANTS AND OUTMIGRANTS FOR BC COMMUNITIES DEPENDENT ON THE FOREST AND OTHER INDUSTRIES

Degree of Forest Sector Dependence	Employment Multiplier	Inmigrant/ Outmigrant Ratio
Forest Specialized	2.15	0.96
Dual	2.64	0.91
Diversified	3.02	1.48
Specialized Non-forest	2.53	1.18
Minor or No Forest	3.26	2.90

Source: Forestry Information Report, Tables 1 and 17, BC-X-314

Table 11

1986 EMPLOYMENT MULTIPLIERS AND RATIO OF INMIGRANTS AND OUTMIGRANTS FOR BC COMMUNITIES DEPENDENT ON THE FOREST AND OTHER INDUSTRIES

Degree of Forest Sector Dependence	Employment Multiplier	Inmigrant/ Outmigrant Ratio
Forest Specialized	2.36	0.82
Dual	3.15	0.71
Diversified	3.36	0.74
Specialized Non-forest	2.98	0.95
Minor or No Forest	3.35	0.72

Source: Forestry Information Report, Tables 2 and 19, BC-X-314

Effectiveness of Intensive Silviculture

Silviculture is the art and science of growing trees by using the agricultural tools of crop science, including genetics, fertilizers, pesticides, pruning, thinning, prescribed fire and replanting. Increased expenditures for silviculture were promoted as a solution to the problem of dwindling timber supplies after the Ministry of Forests acknowledged that its yield prediction system was based on an inevitable decline in future yield. The Ministry's 1984 Forest and Range Analysis stated:

> British Columbia's forests are commonly thought to be managed under a policy of constant production over time. This is not true. Many future second-growth stands will yield smaller harvests at maturity than the existing old growth forests. Application of the sustained yield concept must, therefore, allow for making a transition from using an accumulated inventory of mature timber to relying on annual production from second growth. The general trend is downward to a "sustainable" level unless productivity is enhanced through more intensive management.

This statement conveniently forgets that logging at genuinely sustainable levels (in compliance with a policy of non-declining even flow rate of cut) would have prevented a reduced timber supply. Instead, the policy response from the industry has been to make exaggerated claims about the potential of silviculture to make up for timber supply shortfalls. The 1989 report of the Forest Planning Committee of the Science Council of BC, *A Vision for the Future*, outlines their view of the forest sector as the driving force behind the economy of British Columbia. But its dominance is under assault—not by secondary industry, not by other resource sectors, not by the service sector, but rather by the changing nature of the resource. The Committee goes on to list its objectives, without providing a shred of scientific evidence to support their feasibility:

* To produce, in the short term, a minimal annual harvest of 80 million m³ from the available land base of 30 million hectares.

* To capture the full potential productivity of the same land base so that the growth rate in second-growth forests (whether planted or regenerated naturally) will be increased to 120 million m³ by the year 2020, and to 160 million m³ by the end of the next rotation . . .

Perhaps the extreme enthusiasm was understandable, considering that the forest industry had previously inspired government to increase the cut from 25 million to 75 million cubic metres in the 1960's by changing the utilization standards, lowering the cutting age and adding marginal forest types into the calculations.

A more realistic way to estimate the potential of silviculture is to look at the yields for managed stands for coastal Douglas fir (See Figure 13). The results do not justify the specific claims made by the Forest Planning Committee. Basic silviculture—planting trees—produces an increase over 100 years of less than 10 percent; intensive silviculture—planting and thinning four times—produces less than 25 percent.

In Whose Interest Is BC's Forest Policy?

BC's publicly owned forests are being privatized in a way not unlike the enclosure movement in medieval England in the 1700's. The common denominator then and now is the exercise of power, which can be defined as "the capacity to determine the rules of the game" and "the ability to enforce decisions." When economic power brings with it

Figure 13

THE EFFECT OF SILVICULTURE ON YIELDS OF COASTAL DOUGLAS FIR IN BC AND THE US PACIFIC NORTHWEST OVER A 100-YEAR PERIOD

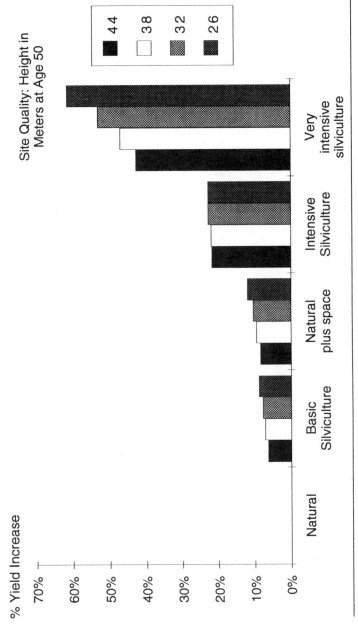

Source: Yield Tables for Managed Stands of Douglas Fir, GTR PNW- 35 (BC and US data)

these capacities, it becomes political power. Certainly economic and political power that reflects the interests of the forest industry functions every day in the workings of BC. In the "free world," power is extended through control of technology, control of capital, and control of production. The economy of British Columbia is planned and organized by such controlling privately owned corporations rather than by the state and its institutions. This is justified by slogans such as "free enterprise" which in theory is supposed to mean "a system in which numerous competing firms strive to sell their goods on an open and unregulated market, each having the same terms for operations as the others, and with government only acting as something of a referee in the market place."

As Patricia Marchak points out in her book *In Whose Interests?* this is the theory, while the Canadian economy actually works like this:

> In reality, the economy is dominated by a very few large corporations which have access unavailable to smaller companies to resources, transportation, and means of marketing goods; and government acts and has always acted as a support for these corporations. There is no actual secret here; indeed . . . a Canadian prime minister (Trudeau) went out of his way to inform Canadians in 1976 "the free market system, in the true sense of the phrase, does not exist in Canada."

For the provincial government to date, good forest management has meant favoring the growth of large integrated resource firms. The process of concentration and integration, which began as early as the 1930s, greatly accelerated in the post-war years with the granting of large forest tenures. This was followed by the move into the BC interior in the 1960s as pulp mill capacity more than doubled. Marchak, in *A History of a Resource Industry*, also notes that the government has favored large companies because

> they are believed to be more reliable (less likely to close down during a recession), more responsible (they have a long term interest in the resource and the labour force), and more profitable (economies of scale produce higher returns to this economy as well as to the producer). In line with these beliefs, governments have channeled public funds towards the provision of an infrastructure of roads, company

towns, and a public service concerned with servicing the industry.

There are obvious fallacies in this argument, partly because of increased out-of-province ownership (See Figure 14). Corporate concentration of the forest industry was addressed in a 1990 BC Central Credit Union newsletter:

> Only 12 years ago, when Premier Bill Bennett declared to the suitors of forest giant MacMillan Bloedel that BC was "not for sale," the province's forest industry was still dominated by locally owned and operated companies. Today, Ontario-based Noranda owns over 50 percent of Mac Blo, and the majority of BC's forest resources are harvested by companies controlled outside the province.
>
> In fact, Eastern Canadian corporations now control 27.1 percent of the assets of BC's largest forest companies (which represent 97 percent of the industry revenues). And, growing from a relatively small base in 1980, foreign-owned multinationals now effectively control 43.2 percent of the assets. Even among the remaining 29.7 percent of the assets, controlled mainly by large local firms, foreign nationals hold minority positions. (For the purpose of this analysis, control is assumed to be ownership in excess of 50 percent. This probably understates the number of companies actually controlled from outside of the province, because effective control can be achieved with much less than half ownership.)
>
> [Foreign companies'] . . . interests do not necessarily coincide with the interests of the people of BC. Individual and multinational shareholders from Toronto, Auckland, New York, Tokyo and Vancouver tend to be more interested in return on investment than in BC's community welfare. They expect management to operate their companies first to produce a profit, and second to be good corporate citizens.

The control of the allowable cut on public forests follows a very similar pattern. In a 1988 *Forest Planning Canada* article, "An Emerging Corporate Nobility" Bill Wagner documented the control that four interlocked groups of companies exercise over 93.2 percent of the allocated forest cut on public lands and 84.1 percent of the overall provincial timber cut on all lands in British Columbia. These four

Figure 14

CONTROL OF BC FOREST COMPANIES BY LOCATION OF MAJORITY SHAREHOLDERS, 1990

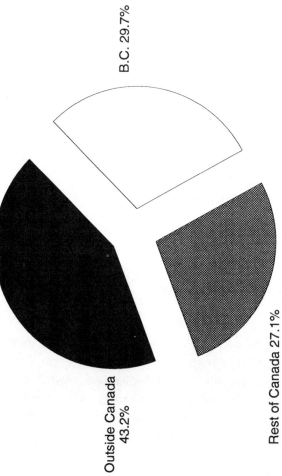

B.C. 29.7%

Outside Canada
43.2%

Rest of Canada 27.1%

Source: Deloitte & Touche (BC Central Credit Union)

groups are the Bentley-Prentice Group, which includes such companies as Takla Forest Products and Canfor Ltd.; the Mead-Scott Group, consisting of such companies as Fletcher Challenge, Whonnock Industries, and Pinnette and Therrien Mills; the Bronfman, Reichman, and Desmarais Group, including Crestbrook Forest Industries and MacMillan Bloedel; and the firms of Sauder, Champion, Ketcham, Fletcher and BCRIC, covering such companies as Weldwood of Canada, West Fraser and Babine Forest Products.

Foreign ownership is a Canadian problem in general. In *Take Back the Nation*, Maude Barlow and Bruce Campbell argue that "Canada is the most economically occupied country in the world; no other country comes close. More than half of our manufacturing sector is under foreign control. The free trade deal with the United States has made this even worse. Rather than becoming export platforms into the new continental market, branch manufacturing plants are simply shutting down."

What Needs Fixing

The key to getting beyond the myths and symbols that dominate BC public policy is to address the issues associated with the forest tenure system and the centralized management structures that support it. To do so will challenge the mechanisms that distribute forest wealth among capital, labour and the resource owners, the public.

The government of British Columbia can no longer ignore the contradictions that exist between public ownership and private use, nor the fact that the potential value of our forests is not being captured. If forest policy is to progress, it must escape from the ideological ghetto in which it now exists, trapped between liberal/pluralist assumptions about the nature of the public policy process and assumptions about the proper role of the state. On the one hand, it pretends the private corporate interest is identical with the public interest. On the other, it pretends that corporate power does not exist and government is free to exercise the public will. Both assumptions are inherent in the forest tenure system and both are false.

British Columbians have seldom questioned the exploitation of natural resources. Perhaps with the high standard of living in BC there has been no incentive to be concerned. But now these internal contradictions are driving dissatisfaction with current forest policy. In April

1991, following two years of hearings and 1700 submissions, the BC Forest Resources Commission released its report, which said:

> It was a daunting task. The forests, the lakes, the rivers, the mountains—they are the heart and soul of the social and economic fabric of this province . . . Thousands talked to us—either directly in the many forums and public meetings held throughout the province, or in written submissions . . . The messages they gave the Commission were as complex and varied as the people themselves, but one theme underlined everything we heard—the status quo is not good enough. The way the forests and their many values are currently being managed by government is out of step with what the public expects. It must change.

Solutions

Policy makers operate in a political context. It is a process that determines who gets what, where and how. In BC, this process has been opportunistic and lacking in accountability. Formally, Richard Simeon in the *Canadian Journal of Political Science* argues the three fundamental dimensions of public policy are the *scope* or the number of areas of economic and social life in which the state is active; the *means* adopted by governments to pursue their policy objectives; and the *distributive* mechanism, which emphasizes the effects of public policy on various groups and classes in society (my emphasis).

Scope

The scope establishes the boundaries of what is considered significant. The timber bias in forestry precludes serious attention to other forest values by limiting its scope. Accordingly, the management system (the means) will have a similarly narrow focus. The existing tenure system therefore distributes benefits to one user group, large forest corporations, while the other groups pay the social, economic and environmental costs. Scope has to take into account non-market and market values of the forest. Humans have many needs. Material needs and values can often be supplied by the market system. Many others cannot. Some forest land must be managed for public goods. The interests of the general public are much broader than the interests of private land owners.

Means

When it comes to means, both public and private ownership is needed to achieve a full range of public and private goods and services, even though public ownership is generally favored by citizens because in theory it provides a much higher degree of control.

Distribution

The distinguishing feature of a public good is that consumption by one individual does not reduce availability of the good for others. The value of a public good is not measured in any market. Examples of public goods are clean air, street lighting, or police protection.

The draft Forest Stewardship Act prepared by a coalition of First Nations, labour and environmental interests in 1991 was an effort to broaden the scope of forest policy, through the means of devolving provincial control to locally accountable Community Forest Boards who are likely to be more motivated to serve a wider range of political interests.

Tenure Options

Rights govern behaviour; they authorize some acts and prohibit others. The search for good government may be viewed as a search for some optimum mix of rights systems. All rights systems have advantages and disadvantages and different comparative advantages in different situations. Crown forest land is a common property resource that all of us own. Certain public goods and services from forest land, such as old growth forest values, will continue only if at least some of the forest is managed as a common property resource.

One of the most fundamental choices government must make is to define property rights, which in turn structures the economic development framework. How rights are defined in forest tenures is the issue in BC forest policy. History has taught British Columbians the folly of former Chief Forester C.D. Orchard's vision of merging public and private rights in the tree farm licence tenures. A priority at this time must be a just and honourable settlement of the land question between First Nations peoples and the federal and provincial governments.

The Australian Shann Turnbull, in his 1980 study *Economic Development of Aboriginal Communities in the Northern Territory*, outlines six broad categories of property rights. There are no distinct boundaries between most of these categories. Forest tenures on Crown land in BC are a blend of Turnbull's groups 1, 2 and 3.

1. **Government, public or social property rights:** This is property owned and controlled by political institutions according to the discretion of individuals who have explicit or implicit power in such institutions. It creates a monolithic political monopoly of the control and use of property, and of power elites in the bureaucracy. Politically, this type of ownership concentrates power and minimizes checks and balances.

2. **Community or collective property rights:** This is the same type of property as the first but on a smaller scale. It retains many of the group 1 characteristics. Like group 1, there are no negotiable shares in the property for its controllers, users and beneficiaries. Non-replaceable forest licences fit into this category.

3. **Contingent or non-discretionary property rights:** These types of property holdings introduce the concept of equity interests on a contingent and conditional basis. The equity interest is contingent upon some type of contribution by those who receive the benefits under specified terms and conditions. Tree farm licences and forest licences fit into this category. These licences have a value and windfall profits have been realized in their sale.

4. **Capitalistic or discretionary property rights:** This is property held in fee simple, where the owners can negotiate, sell or otherwise use the property at their discretion. Crown grants are in this category.

5. **Dynamic tenure rights:** This tenure concept was developed by Shann Turnbull. The rights and obligations of the tenure relationship are designed to change over time. For instance, a tenure can be designed to enable the investor to recover his investment in the initial period while escalating the equity interests over a period of (say) 15 to 25 years to eventually allow for 100 percent resident ownership and control.

6. **Aboriginal tenure rights:** These rights are primarily concerned with the relationships between aboriginal people and land. The basic determinant is the family heritage. Aboriginals considers themselves part of the land, and the relationship is more of being an "ownee" rather than an "owner."

The broad outline of the options for tenure reform are contained in Marion Clawson's essay "Major Alternatives for the Future Management of the Federal Lands" in *Rethinking the Federal Lands*, which have been modified for BC:

1. **Most or all of the present public forests could be retained in provincial ownership,** but major efforts should be made to improve their management, with more concern for the costs and returns from such ownership and management.

2. **Most or all present public forests could be turned over to Community Forest Boards** without charge or on payment of some price, with their future management or disposal to be determined by the Board.

3. **A major part of the present provincial land could be sold** to private individuals, corporations, and groups or associations under terms to be spelled out in the enabling legislation.

4. **All or large parts of the present provincial forest land area could be transferred to public or mixed public-private corporations,** to be managed as decided upon by those corporations, but under terms of the enabling legislation.

5. **Long-term leasing of provincial lands could be extended,** not only as it has been done in existing tenures but also for other commercial uses, and for conservation or preservation purposes as well. Small forest tenures for woodlots are in this category.

Land use policy reform in British Columbia has stagnated for too many years. The time has come for a wide public discussion of the rights systems and tenure options available for structuring the kind of society we want to be. For a new policy to emerge, a primary determinant will be a consensus on what the proper role of government should be in forest management. It is time to think big and to think long term.

Although the specifics of what is needed are not known, we do know enough to change. Our provincial forests are a magnificent asset and too important not to be managed efficiently for the greatest social and economic good. The inefficiencies of present policy are far too large. It is time to innovate.

Sources
Apsey, Mike. 1991. "What The Forest Industry Wants." *Globe and Mail*, February 28, 1991.

Associated Press. 1992. "Canada Losing Competitive Edge as Japan Holds Top Spot—Report." *Times Colonist*. June 23, 1992.

Barlow, Maude and Bruce Campbell. 1991. *Take Back the Nation*. Key Porter Books.

Baxter-Moore, Nicolas. 1987. "Policy Implementation and the Role of the State: A Revised Approach to the Study of Policy Instruments." In *Contemporary Canadian Politics*. Prentice-Hall Canada Ltd.

BC Central Credit Union. 1990. "A Tale of Two Economies." *Economic Analysis of British Columbia*. Volume 10(2).

— 1990. "Who Controls BC's Forest Industry." Volume 10(6).

Brubaker, Sterling, editor. 1984. *Rethinking the Federal Lands*. Resources for the Future Inc. Washington DC.

Cail, Robert E. 1974. *Man, Land, and the Law: The Disposal of Crown Lands in British Columbia 1871-1913*. University of British Columbia Press.

Canadian Employment and Immigration Advisory Council. 1987. *Canada's Single-Industry Communities: A Proud Determination to Survive*. Report to the Minister, Government of Canada.

Canadian Labour Market and Productivity Center. 1992. *Environmental Protection and Jobs in Canada: A Discussion Paper for Business and Labour*. Government of Canada.

Canadian Press. 1990. "Canada Becoming Less Competitive." *Times Colonist*. June 20.

Charlie, Chilcotin. 1992. "MOF Responds." *Big Creek Bugle*.

Clawson, Marion. 1976. *The Economics of National Forest Management*. Resources for the Future, Washington, DC.

—1984. "Major Alternatives for the Future Management of the Federal Lands." *Rethinking the Federal Lands*. Sterling Brubaker, ed. Resources for the Future.

Edelman, Murray. 1964. *Politics as Symbolic Action: Mass Arousal and Quiescence*. Academic Press.

—1977. *The Symbolic Uses of Politics*. University of Chicago Press.

Elton, Bob. 1992. *Financial Condition of the Industry*. British Columbia Forest Industry Conference. Price Waterhouse.

Fisher, Robin. 1977. *Contact and Conflict: Indian-European Relations in British Columbia, 1774-1890*. University of British Columbia Press.

Forest Planning Committee. 1989. *A Vision for the Future*. The Science Council of BC.

Forest Resources Commission. 1991. *The Future of Our Forests*. Province of British Columbia.

1992. *British Columbia Employment Dependencies*. Province of British

Columbia.

Gough, Barry M. 1984. *Gunboat Frontier: British Maritime Authority and Northwest Coast Indians, 1846-90.* University of British Columbia Press.

Mahood, Ian and Ken Drushka. 1990. *Three Men and a Forester.* Harbour Publishing.

Marchak, Patricia M. 1975. *Ideological Perspectives in Canada.* McGraw-Hill Ryerson Ltd.

—1979. *In Whose Interests? An Essay on Multinationals in Canada.* McClelland and Stewart.

—1989. "History of a Resource Industry." *A History of British Columbia.* Copp Clark Pitman Ltd.

Ministry of Forests. 1984. *Forest and Range Resource Analysis.* Province of British Columbia.

Mulholland, F.D. 1939. "Forest Policy—Ownership and Administration." *Forestry Chronicle.*

—1947. "The Relation Between the State and Private Forestry." *Forestry Chronicle.* Volume 23, pp 29-35.

Newman, Peter C. 1988. "Caesars of the Wilderness." *Company of Adventurers.* Penguin Books.

Nixon, Bob. 1990. "The Death of Another Forestry Myth: Fifty Cents of Every Dollar." *Forest Planning Canada.* Volume 6(4):14.

Orchard, C.D. No Date. *Reminiscences.* Box 4, File 20, Orchard Papers.

—1942. *Forest Working Circles.* Box 8, File 15, Orchard Papers.

—1946. "The Function of the State in the Management of Crown and Private Forests for the Production of an Assured Supply of Wood for Industry." *Forestry Chronicle.* Volume 22, pp 100-109.

Roach, Thomas R. 1984. "Stewards of the People's Wealth: The Founding of British Columbia's Forest Branch." *Journal of Forest History.* January.

Roy, Patricia E. 1989. *A History of British Columbia.* Selected Readings. Copp Clark Pitman Ltd.

Simeon, Richard. 1976. "Studying Public Policy." *Canadian Journal of Political Science.* Volume IX (4): 550- 580.

Sloan, Hon. Gordon McG. 1956. *The Forest Resources of British Columbia.* Report of the Commissioner. Volume 1 and 2. Province of British Columbia.

Spring, Joel H. 1972. *Education and the Rise of the Corporate State.* Beacon Press.

Telford, R.C. "Sustained Production in British Columbia." *Forestry Chronicle.* Volume 24, pp. 22-26.

Tin Wis Coalition. 1991. *Draft Forest Stewardship Act.*

Turnbull, Shann. 1980. *Economic Development of Aboriginal Communities in the Northern Territory*. Department of Aboriginal Affairs. Australian Government Publishing Service.

Vickers, Sir Geoffrey. 1965. *The Art of Judgment: A Study of Policy Making*. Chapman and Hall.

Wagner, Bill. 1988. "An Emerging Corporate Nobility? Industrial Concentration of Economic Power on Public Timber Tenures." *Forest Planning Canada*. Volume 4(2):14- 19.

Westoby, Jack. 1987. *The Purpose of Forests: Follies of Development*. Basil Blackwell.

White, W.A., K.M. Duke and K. Fong. 1989. *The Influence of Forest Sector Dependence on the Socio-economic Characteristics of Rural British Columbia*. Forestry Canada Information Report BC-X-314.

Whitehead, Alfred North. 1933. *Adventures of Ideas*. The MacMillan Company.

Wilson, Jeremy. 1986. *Forest Conservation in British Columbia, 1935-85*: Reflections on a Barren Political Debate. For Presentation at the Fourth BC Studies Conference.

Worrell, Albert C. 1970. *Principles of Forest Policy*. McGraw-Hill Book Company.

Contributors

Ken Drushka is a Vancouver writer and editor specializing in forestry issues. He worked for sixteen years logging, silvicultural contracting and sawmilling. He has written several books on the BC forest industry, including *Working in the Woods*, published in 1992.

Bob Nixon, a professional forester (Society of American Foresters), moved to BC in 1977 to work as conservation representative for the Sierra Club of Western Canada. In 1985 he launched *Forest Planning Canada*, a monthly in-depth journal on forestry issues which he still publishes.

Dr. M. Patricia Marchak is Dean of Arts and a sociology professor at the University of British Columbia, Vancouver, BC. She represented the New Democratic Party in Vancouver–Point Grey in the 1983 provincial election, and she has written numerous books and articles on forestry and other issues.

Julian Dunster is a Registered Professional Forester, Professional Planner and Certified Arborist. His work as a consultant, lecturer and writer on urban forestry and environmental impact issues has taken him across Canada and to Britain, Scandinavia, Australia, New Zealand and Nepal.

Herb Hammond, author of *Seeing the Forest Among the Trees*, is a Registered Professional Forester, teacher, forest ecologist and citizen activist. Specializing in wholistic forest use, he works with First Nations, rural water users, environmental groups and communities across Canada.

Holly Nathan grew up in Port Alberni, BC, where she attended Smith School with Ahousat students. She has served as Native affairs reporter for the Victoria *Times Colonist* for six years.

Ray Travers, a Registered Professional Forester, has served with private industry and governments in BC and the Northwest Territories. Currently he works as a consultant specializing in forest policy analysis, resource planning, community forestry and environmental impact assessment.

Index

marbled murrelet, 26
Marchak, M. Patricia, 67–84, 172, 174, 214
Marubini, 75
Masset, 167
McCloskey, Kelly, 150
McEachern, Allen, 154, 166
McLeod Lake Indian Band, 167
Mead–Scott Group, 217
Meares Island
    (Wah-nah-juss/Hilth-hoo-iss), 48, 51, 151, 156, 165, 166
Meares Island Planning Team, 64
Memorandum of Understanding on
    Softwood Lumber, 200
Merkle, Gary, 152
Mewitty, 176
Mexico, 72
microorganisms, 103, 104, 112, 118, 121, 134
Miller, Dan, VIII, 28, 30, 33, 36, 37, 50, 154
Minas Geras, 74
Ministry of Agriculture and Fisheries
    (BC), 33, 41
Ministry of Environment (BC), 24, 25, 44, 153, 154
Ministry of Forests (BC), 15, 24–26, 27, 28, 29, 36, 40, 42–45, 48, 50, 52, 55, 56, 57, 59, 99, 100, 109, 116, 167, 171, 185, 192, 197, 200, 201, 211
Ministry of Natural Resources
    (Ontario), 87, 90, 91
Mission, 20
Mitsubishi, 74
Model Forests, 149
Monk, Justa, 160
Monterey pine *see* radiata pine
Moore, Harry, 157
moose, 119
Morocco, 70
Morrison, Michael L., 121
Mount Paxton, 139
Mulholland, F.D., XI, 7, 8, 179, 180

Mulroney, Brian, 26, 35, 38, 59
mushroom, 162, 163, 165
Musqueam Nation, 137
mycorrhizal fungi, 112, 113

N
Nanaimo, 24, 140
Nass River Valley, 139
Nathan, Holly, 137–170
National Congress on Natural
    Resources (USA), 178
National Environmental Policy Act
    (NEPA), 88, 95
National Forest Management Act
    (NFMA), 89
National Geographic, 139
Native bands, 17
Native communities, 29
Nelson, 20, 200
Nelson, Robert, 173
Nemiah Valley Indian Band, 166
New Brunswick, 196, 198
New Democratic Party, VII, 21, 28, 37, 39, 50, 63, 138, 153, 157, 158, 172, 173
New Westminster, 176
New Zealand, 68, 73, 74, 76, 77, 196
New Zealand Forest Products, 76
Newfoundland, 196, 198
Newman, Peter C., 176
Nicola Valley Tribal Council, 144
Niku, 80
Nimpkish band, 164
Nisga'a, 138, 139
Nixon, Bob, 23–66, 182
Noranda, 78, 149, 215
Nordic countries, 79, 81
North America, 18, 23, 67, 73, 74, 98
North American Paper Group (Reed
    International), 75
North Broken Hill Peko, 78
North Cowichan, 20
northern spotted owl, 13
Northwood Pulp Company, 149

Watts, Richard, 163
Weldwood, 75, 217
Wells Gray, A., xi, 1
Wesley Vale, 78
West Chilcotin Community Resources
    Board, 46, 49, 42, 108
western yew, 143
West Fraser, 75, 217
West Moberly Band, 162
Westshore Terminals Ltd., 154
Weyerhaeuser, 7, 78
whales, 103
wheat, 72
White, W.A., 209
Whitehead, Alfred North, 171
Whitford, H.N., xi
Whonnock Industries, 217

Wilderness Advisory Committee, 56
Williams, Bob, 172
Williams, Violet, 143
Williams Lake, 109, 200
Wilson, Jeremy, 182
Wondolleck, J.M., 93
Woodlot Licence, 14, 17, 19, 139, 149
Wood Supply Agreements, 17
World Economic Forum, 183
World War Two, 6

XYZ
Xax'lip Nation, 150, 153
Yoos, 141, 142
Young, Bill, 48
Yuen Foong Yu, 76
Zeballos, 145